Pipeline Engineering Monograph Series

Pipeline Geo-Environmental Design and Geohazard Management

Edited by Moness Rizkalla

Library of Congress Cataloging-in-Publication Data

Rizkalla, Moness.
  Pipeline geo-environmental design and geohazard management / by Moness
Rizkalla.
     p. cm.
  Includes bibliographical references and index.
  ISBN 978-0-7918-0281-6
  1. Pipelines--Design and construction. 2. Environmental protection. I. Title.

 TJ930.R59 2008
 621.8'6720286--dc22
                                    2008032683

Table of Contents

CHAPTER 1. Introduction   **1**
   By Moness Rizkalla

CHAPTER 2. Pipeline Route Selection and   **5**
Characterization
   By J.D. Mollard, Lynden Penner and Troy Zimmer

   2.1   Route Selection and Characterization   5
   2.2   Integrating Multidiscipline Study Data   8
   2.3   Remote Sensing Pipeline Routes   9
   from Maps, Satellite Imagery and Airphotos
   2.4   Nine Global Terrains and   22
   Associated Geohazards
   2.5   Synthesizing Terrain Datasets in a GIS   58
   2.6   Presenting pipeline and Terrain Data   60
   in a GIS
         Acknowledgments   62
         References and Additional Reading   63

CHAPTER 3. Open Cut and Elevated River   **69**
Crossings
   By Wim Veldman

   3.1   Introduction   69
   3.2   Design   70
   3.3   Construction   111
   3.4   Operations   126
   3.5   References   132

CHAPTER 4. Horizontal Directional Drilling   **133**
   By J. D. Hair and Jeffrey Puckett

   4.1   Introduction   133
   4.2   The Horizontal Directional Drilling   133
   Process

4.3 Site Investigation 139
4.4 Drill Path Design 140
4.5 Temporary Workspace Requirements 142
4.6 Drilling Fluids 145
4.7 Pipe Specification 151
4.8 Contractual Considerations 163
4.9 Construction Monitoring 164
References 166

CHAPTER 5. Buoyancy 169
By Ray Boivin

5.1 Introduction 169
5.2 Definitions and Abbreviations 170
5.3 Pipeline Codes 172
5.4 Buoyancy Design Philosophy 173
5.5 Buoyancy Control Options 174
5.6 Buoyancy Design Forces 207
5.7 Buoyancy Control Spacing and Locations 214
5.8 References 226

CHAPTER 6. Geohazard Management 229
By Moness Rizkalla, Rodney Read and Gregg O'Neil

6.1 Introduction 229
6.2 Regional Geohazards 233
6.3 Data Management 244
6.4 Risk Assessment Methodologies 253
6.5 Monitoring 285
6.6 Design and Operational Mitigation 308
6.7 Geohazard Management Planning Process 308
6.8 References and Related Readings 334

About the Contributors 351

Preface to Pipeline Engineering Monograph Series

The editorial board of the ASME Pipeline Engineering Monograph series seeks to cover various facets of pipeline engineering. This monograph series puts emphasis on practical applications and current practices in the pipeline industry. Each book is intended to enhance the learning process for pipeline engineering students and to provide authoritative references of current pipeline engineering practices for practicing engineers.

Pipeline engineering information is neither readily available from a single source nor covered comprehensively in a single volume. Additionally, many pipeline engineers have acquired their knowledge through on-the-job training together with short courses or seminars. On-the-job training may not be comprehensive and courses or seminars tend to be oriented toward specific functional areas and tasks.

The editorial board has tried to compile a comprehensive collection of relevant pipeline engineering information in this series. The books in this monograph series were written to fill the gap between the basic engineering principles learned from the academic world and the solutions that may be applied to practical pipeline engineering problems. The purpose of these books is to show how pipeline engineering concepts and techniques can be applied to solve the problems with which engineers are confronted and to provide them with the knowledge they need in order to make informed decisions.

The editorial board has sought to present the material so that practicing engineers and graduate level pipeline engineering students may easily understand it. Although the monograph contains introductory material from a pipeline engineering viewpoint, it is reasonably comprehensive and requires a basic understanding of undergraduate engineering subjects. For example, students or engineers need to have basic knowledge of material corrosion mechanisms in order to understand pipe corrosion.

Each book or chapter starts with engineering fundamentals to establish a clear understanding of the engineering principles and theories. These are followed by a discussion of the latest practices in the pipeline industry, and if necessary, new emerging technologies even if they are not as yet widely practiced. Controversial techniques may be identified, but not construed as a recommendation. Examples are included where appropriate to aid the reader in gaining a working knowledge of the material. For a more in-depth treatment of advanced topics, technical papers are included. The monographs in this series may be published in various forms; some in complete text form, some as a collection of key papers published in journals or conference proceedings, or some as a combinations of both.

The editorial board plans to publish the following pipeline engineering topics:

- Pipe Material
- Pipeline Corrosion
- Pipeline Integrity
- Pipeline Inspection
- Pipeline Risk Management
- Pipeline System Automation and Control
- Pipeline System Design
- Pipeline Geo-Environmental Design and Geohazard Management
- Pipeline Project Management
- Pipeline Codes and Standards

Other topics may be added to the series at the recommendation of the users and at the discretion of the editorial board.

The books in this monograph series will be of considerable help to pipeline engineering students and practicing engineers. The editorial board hopes that pipeline engineers can gain expert knowledge and save an immeasurable amount of time through use of these books.

Acknowledgments

We, on the editorial board, wish to express our sincere gratitude to the authors, editors and reviewers for their great contributions. They managed each volume, wrote technical sections, offered many ideas, and contributed valuable suggestions. Financial support from the Pipeline Systems Division (PSD) of ASME enabled us to create this monograph series, providing the crucial remainder to the time and expenses already incurred by the editors and authors themselves. We are indebted to the organizing and technical committees of the International Pipeline Conferences (IPC), which have provided an excellent forum to share pipeline engineering expertise throughout the international pipeline community. We were fortunate to have the skillful assistance of the publication department of ASME not only to publish this series but also to undertake this non-trivial task.

Editorial Board

1 Introduction

From a pipeline operator's viewpoint, the common elements between the aspects of pipeline geo-environmental design and geohazard management discussed in this book are the pipeline company's responsibility to protect the environment in the vicinity of the pipeline and its business interest to protect the pipeline from potentially adverse environmental conditions. The daily work in pipeline companies in the topics addressed in this book is executed in the multi-disciplinary space shared by project managers, pipeline engineers, construction staff and both geotechnical and hydrotechnical specialists.

Five widely applicable and critically important topics to pipeline designers and operators were selected to be addressed by recognized specialists in their respective fields. Brief biographic summaries of the invited contributing authors are included at the end of the book. The structure of the book and the contributing authors is as follows:

- Chapter 2: Pipeline Route Selection and Characterization by Dr. Jack Mollard, Mr. Lynden Penner and Mr. Troy Zimmer
- Chapter 3: Open Cut and Elevated Pipeline River Crossings by Mr. Wim Veldman
- Chapter 4: Horizontal Directional Drilling by Mr. John Hair and Mr. Jeffrey Puckett
- Chapter 5: Overland Buoyancy Control by Mr. Ray Boivin
- Chapter 6: Geohazard Management by Mr. Moness Rizkalla, Dr. Rodney Read and Mr. Gregg O'Neil

In recent years, there have been valuable experiences gained by the pipeline engineering and construction community as a result of having to deal with challenging environments in the routing, design and construction of recently completed major pipeline projects as well as in meeting the heightened expectations of ensuring the integrity of aging operating systems. As a result, the topics addressed in this book have seen significant advances in recent years that have redefined the state-of-practice in how to design, construct and ensure the integrity of operating pipelines.

This book in is intended to provide a working knowledge for non-specialists so as to better define the main issues during design, construction and integrity management planning during operations. To meet its objective, most chapters include many photographs and figures of practical applications from projects world wide. Additionally, for 2 chapters, the authors elected to include additional references both for completeness and as recognition of the considerable work by others.

Recognizing the multi-disciplinary nature of optimizing a pipeline's alignment during the early planning and design stages, chapter 2 presents an overview of pipeline route selection and characterization. During that stage of pipeline development, a balance is required in addressing engineering, biophysical and socioeconomic factors. The integration of multi-disciplinary data sets identifies boundaries for routing to address a pipeline proponent's range of responsibilities to land owners, land users and to the environment and wildlife. The extensive use of remote sensing products including maps, satellite imagery and air photography is a mark of this stage of a project's development. An extensive discussion is offered of nine global terrain types to serve as an input of what may be expected for common ground conditions by the route selection team and the pipeline's designers and construction management team. Finally, the now ubiquitous application of Geographic Information Systems (GIS) in synthesizing all data sets to support the decision making and communication of pipeline routing is briefly discussed. For the benefit of the readers, the authors of this chapter elected to include additional related readings beyond those used explicitly as references

The importance of water crossing site selection, design, construction and integrity management during operations is reflected in the discussions of chapters 3 and 4. Chapter 3 addresses open cut and elevated pipeline river crossings. Chapter 4 discusses horizontal directional drilling (HDD)

The discussion of open cut and elevated pipeline river crossings in chapter 3 starts with an overview of design considerations and includes design examples. Most importantly, guidance is provided in terms of what is critical and what is less critical in terms of design inputs. A discussion of construction methods follows. Finally, a brief but valuable discussion is presented on recommended monitoring during operations and of lessons learned over years of observations of arctic river crossings.

The discussion of HDD in chapter 4 begins with a description of the pipeline installation process starting with a pilot hole, followed by pre-reaming and through to pull-back of the product pipeline. A discussion of the necessary site investigation is offered before describing the drill path design. Temporary workspace requirements, drilling fluids and pipeline specifications are addressed in turn. Given the complexity and the commonly sub-contracted arrangement of HDD installations, an overview of contractual considerations is included. Finally, and following through to the completion of the installation, guidance is provided on monitoring during construction.

Buoyancy control is another geo-environmental design and construction topic that is addressed in this book as chapter 5. A full menu of buoyancy control option is presented ranging from the variety of concrete weights, to soil weight based methods to a variety of anchors. The factors of buoyancy control design are presented and various input values are suggested. The design considerations addressed are: backfill shear resistance, density values of water, design safety

2

factors, soil liquefaction and both pipeline stress and operating temperature. Buoyancy control spacing calculations and a comparison between the various methods is presented. Finally, an overview is offered of the practical process of the design, procurement and installation of these pipeline design elements.

The sixth and final chapter of the book addresses geohazard management. This too is an area that has seen considerable advancements in recent years as the topic has gained prominence within the pipeline community. Forty geohazards in 10 broad categories that can occur in various parts of the world are identified and their potential impacts on the pipeline itself, the ditch or the right-of-way are noted. An overview is then presented of the data management aspects of this strongly data-dependent topic. A fuller discussion follows of the range of risk assessment methods including examples from within the pipeline industry. Overviews of the ranges of both geohazard monitoring and mitigation technologies methods are presented. Finally, an integrated geohazard management planning process is offered for practical application by operating pipeline companies throughout the lifecycle of a pipeline. For the benefit of the readers, the authors of this chapter elected to include additional related readings beyond those used explicitly as references.

Every care has been exercised by the authors to contact copyright holders and obtain permissions and reference materials and to avoid errors and omissions. Notifications of corrections, omissions and attributions will be welcomed by the authors.

2   Pipeline Route Selection and Characterization

2.1 Route Selection and Characterization

Pipeline route selection is the process that evaluates and selects a preferred route from competing alternative routes. The process is largely governed by such controls as engineering requirements and costs, environmental (biophysical) concerns and socioeconomic issues (Figure 1). Pipeline route selection also requires consultations with government agencies, the public, affected communities and multi-agency regulators. Each of these controls influences some aspect of pipeline route selection.

Pipeline route characterization is the process that determines the nature and behaviour of factors that influence and are influenced by the pipeline route. In the overall route selection process, route characterization addresses the combined effects of engineering and environmental factors as well as socioeconomic and political controls (Figure 1). This chapter deals extensively with characterization of terrain conditions and geohazards that affect pipeline route selection and pipeline construction, maintenance and operation in nine distinctive global terrains.

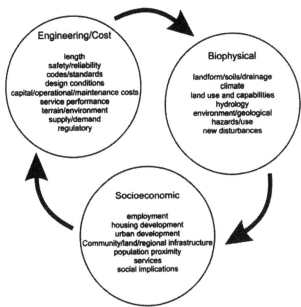

Figure 1 – Some engineering, cost, biophysical and socioeconomic factors influencing pipeline route selection and characterization
(*from* **Mohitpour, *et al.*, 1998**)

The characterization of terrain and natural geohazards requires the collection and interpretation of information on land use and land cover, topography, geomorphic features, geology, hydrogeology, hydrology, hydrography, seismology and climate, all of which affect the geotechnical, hydrotechnical and geoenvironmental characteristics of terrain and its suitability for a pipeline route. The emphasis in this chapter is on interpreting remotely sensed information to identify geohazards. In pipeline route selection and characterization studies, remote sensing analysts use airphotos and satellite imagery to identify Earth surface features and conditions and the genetic processes and environments that created them, and furthermore to infer their geoengineering and geoenvironmental characteristics. Office remote sensing studies reveal difficult terrains and geohazards that can be avoided in initial route selection and guide follow-up site-specific field investigations along competing and selected pipeline routes (Figures 2a and 2b).

But pipeline route selection is also guided by the pipeline owner's needs and requirements. As a result, the route selection geoengineering/geoenvironmental team works closely with the project proponent and the proponent's principal pipeline construction engineer as well as with other disciplines contributing to the selection of a final preferred pipeline route.

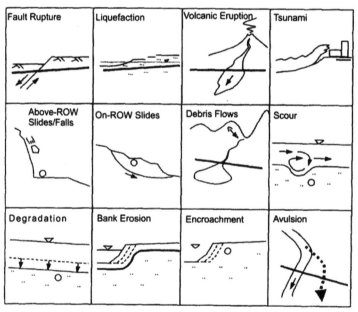

Figure 2a – Twelve cartoons illustrating pipeline geohazards
(*from* **Porter** *et al.*, **2005**)

6

Geohazard	Description	Operation & Maintenance Risks	Geotechnical Mitigation Options		
			Investigation	Routing	Design & Construction
Earthquakes - fault rupture	Movement likely along pre-existing fault lines during earthquake activity	Displacement, deformation, rupture, uncontrolled spillage. Can be trigger for landslides and ground collapse.	Locate fault zones. Assess earthquake history	Detailed alignment control in fault zones	Special trench design.
Earthquakes - liquefaction	Ground shaking causes liquefaction of loose fine granular and metastable soils	Loss of support, displacement, deformation, rupture, uncontrolled spillage	Locate and classify material types and soil structure	Avoid susceptible soils by lateral realignment or deepening.	Soil improvement
Volcanoes	Dome eruption, lava flows, ejected material, lahars	Displacement, deformation, rupture, loading, uncontrolled spillage	Locate existing domes and flow channels	Avoid + local alignment control to negotiate existing flow channels	
Landslides	Slow or rapid ground displacement caused by change in geometry, groundwater level or seismicity, includes rock fall, shallow soil slides, deep rotational slides, debris and mud flows to significant distance from source	Loss of support, displacement, deformation, rupture, loading, uncontrolled spillage	Locate existing landslides, landslide prone terrain and extent of sidelong ground	Detailed alignment control to avoid existing landslides. Minimise potentially unstable sidelong ground.	Careful earthworks design, spoil handling measures and reinstatement
Gullying and Soil Erosion	Removal of soil by water action across and adjacent to the pipeline. Existing gullies prone to enlargement by erosion and scour of banks and headwall.	Loss of support, displacement, deformation, rupture, uncontrolled spillage	Assess extent of sidelong ground, locate gullies and assess catchment	Minimise sidelong ground and avoid areas of active erosion.	Careful design of earthworks, cross drainage and erosion protection measures.
Karst-Limestone	Limestone that has been or continues to dissolve in groundwater resulting in a network of sink holes, caves, etc. Prone to sudden collapse.	Loss of support, displacement, deformation, rupture, uncontrolled spillage into groundwater system	Assess and classify extent of karstification	Avoid, minimise, detailed alignment control	Ground improvement measures
Karst-Gypsum	Gypsum or other sulphate enriched soils and rocks that have been or continue to dissolve in groundwater resulting in a network of sink holes, caves, etc. Prone to sudden collapse.	Loss of support, displacement, deformation, rupture, uncontrolled spillage into groundwater system	Assess and classify extent of karstification	Avoid, minimise, detailed alignment control	Ground improvement measures
River Channel Migration	River channels migrate across wide valley floors and sudden changes in location can occur under flood conditions	Loss of support, displacement, deformation, rupture, uncontrolled spillage into river system	Map valley floor. Assess catchment and hydrological history	Minimise crossing length	Pipe bridges or maintain sufficient depth of burial, river control

Figure 2b – Terrain constraints and geohazards affecting pipeline route selection (*from* Charman *et al.*, 2005)

2.2 Integrating Multidiscipline Study Data

To begin the route selection process, the geoengineering/geoenvironmental team requires preliminary conceptual information from the pipeline proponent specifying the pipeline origin, destination, diameter, product to be carried, operational specifications, and so on. The origin of pipeline routes is commonly determined by the location of an oil or gas field and the associated pipeline gathering system. The pipeline destination location, on the other hand, may be the product customer, a processing facility or a connecting pipeline system. Fixed points at compressors and pump station facilities often serve as intermediate route location control points.

During initial conceptual planning, pipeline routing engineers and geoscientists obtain route directions from the proponent's senior management. They also get preliminary information on potential route corridors from the biophysical, environmental, socioeconomic, political and cultural route selection teams. All teams require information from various levels of government, the general public and local communities affected by the pipeline. Some of the important data the teams need to collect include governmental multi-agency regulations, the concerns of stakeholders and route intervenor issues. Because these input data significantly affect the selection of the final route, it is important that the various tasks of all teams be properly planned, co-ordinated and integrated at the start of the route selection process.

One of the first tasks of the geoengineering team is to assemble a database of geological, geotechnical, geophysical, hydrotechnical, geoenvironmental and geohazard information that guides pipeline route selection and characterization. The environmental team, on the other hand, identifies and contributes information on protected lands, environmentally sensitive wildlife, fish and bird habitats, endangered species and archaeological sites (Figure 3). These data are required to conduct the environmental impact assessment (EIA) for pipeline project approval. The input of socioeconomic, cultural and government public consultation teams also has a strong influence on the final pipeline route. They collect viewpoints and apprehensions from a variety of supporting and opposing stakeholders—from those hoping to secure jobs for local workers and from others objecting to the project.

Following preparatory meetings with various team members, and taking into account any initial restrictions placed on pipeline route selection, members of the geoengineering/geoenvironmental team begin the interpretation of remotely sensed terrain and geohazard data from multidisciplinary maps, satellite imagery and airphotos.

- National, provincial and state parks
- National and provincial reserves, forests and monuments
- Aboriginal lands
- Bird sanctuaries
- Fish habitat
- Special interest areas
- Natural and historic features
- Federal crown land
- Special conservation areas
- Archaeological sites
- Endangered species: plants, animals, birds

Figure 3 – Environmental constraints affecting pipeline route selection
(*from* **Charman** *et al.*, **2005**)

2.3 Remote Sensing Pipeline Routes from Maps, Satellite Imagery and Airphotos

A great many difficult pipeline construction terrains and geohazards can be discovered early in the route selection process by skilled engineers and geoscientists interpreting remotely sensed maps, satellite imagery and airphotos, thereby avoiding or minimizing difficult and hazardous terrains. Where a costly or hazardous route selection problem can't be avoided, remotely sensed terrain information is used to direct field personnel to specific sites to investigate the identified problem. Remote sensing is an essential tool for developing a reliable appreciation of terrain types and ground conditions, and their origin, behavior and significance—not only along competing route corridors but also on narrowly constrained adjustments within these corridors, as was the case for a pipeline route selection study from Prudhoe Bay, Alaska, to Zama, northwestern Alberta (Figure 4).

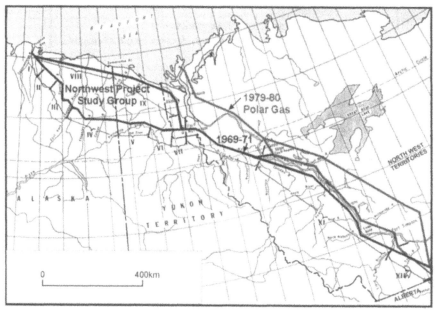

Figure 4 – Northern pipeline route selection and terrain mapping from airphoto analysis and field reconnaissance, 1969 to 1980
(from, Mollard, 1971)

2.3.1   Acquisition of Maps, Satellite and Airphoto Imagery

Multidiscipline maps, satellite images and airphotos can usually be acquired from federal, state, provincial, territorial and local government agencies. Airphotos and satellite imagery can also be obtained from national and international commercial agencies. Depending on available resources and difficulty of the route, it may be cost effective to contract new aerial photography. An increasing number and variety of satellite imagery types, spatial resolutions and scales (Figure 5) are available from national and international government agencies and private sources—often at small cost relative to potential savings.

Sensor Type	Spatial Resolution	Minimum feature size (m)	Stereo Coverage
Landsat 7 ETM+	30m (multispectral) 15m (panchromatic)	45	No
SPOT IV	20m (multispectral) 10m (panchromatic)	30	Yes
IKONOS	4m (multispectral) 1m panchromatic)	12 3	Yes
IRS-1D	23m (multispectral) 6m (panchromatic)	72 18	No
Radarsat	8m	Variable	No
ERS-1	25m	Variable	No
Quickbird	2.4m (multispectral) 0.6m (panchromatic)	1.8	Yes
ENVISAT (forthcoming)	30m	Variable	No
Orbview 3 (forthcoming	4m (multispectral) 1m (panchromatic)	3	No

Figure 5 – Satellite sensor types (*modified after* Hearn, 2005)

In most cases, the single most useful data source for pipeline route selection is black-and-white stereoscopic aerial photography. Stereoscopic airphotos covering the pipeline corridor can be selected from flight-index sheets available on the Internet for many jurisdictions. Desired airphotos can be ordered by fax or email from national, state and provincial airphoto libraries, or by phoning directly to airphoto library staff. While the most common and readily available conventional airphoto is black-and-white panchromatic, some libraries have true color, black-and-white and false color infrared airphotos on computer files as well as several different ages and scales of conventional airphotos. As a result, those looking for terrain geohazards and significant changes in the natural and cultural landscape can select from a large menu of airphotos. Moreover, the airphotos can be orthorectified for digital elevation modelling.

Aerial LIDAR (Light Detection and Ranging) imagery, acquired from low-flying aircraft, is a valuable remote sensing tool for topographic mapping. LIDAR detects and measures both the top of the vegetation canopy and the ground surface below. It is highly accurate (±50 mm) and can be rapidly acquired. Therefore, it is well-suited to long linear pipeline routes over a variety of terrain types and conditions.

2.3.2 Map Remote Sensing

Interpreting map data is an economical form of terrain and geohazard information remote sensing. Maps provide a low-cost, readily available collection of data on soils, geology, topography and other types of physical and human geography data that can guide, support and supplement airphoto and satellite imagery analysis.

The list of maps can include physiographic, topographic, cadastral, political, climatic, surficial and bedrock geology, agricultural soil survey, hydrogeology, bathymetry, land use and land cover. And they may be combined digitally to assess the integrated effects of pipeline construction and operation, as, for example, on soil erosion (Figure 6). Figure 7 is a further illustration that combines a larger group of terrain, environmental, socioeconomic and political factors affecting pipeline route selection. Today, many maps are available in digital and traditional paper format. Depending on the region or local pipeline area under observation, the list may include maps of mines and minerals; national, provincial or state parks; wildlife refuges, preserves and sanctuaries; biophysical and ecological maps providing environmental data on fish, birds, mammals, endangered flora and fauna species, wilderness and wildland areas, vegetation and forestry, archaeological sites and aboriginal reserves. These maps often show protected lands that limit pipeline route options, and are best examined during the early stages of pipeline route selection. Other maps show existing transportation and communication infrastructure corridors and networks that can impact pipeline route selection. The cost of these maps is inconsequential compared to the overall pipeline project cost, and they, along with contact airphotos and satellite images, should be obtained early in the project planning phase of route selection.

Depending on the time and purpose for which a multidiscipline map was made, it may not be detailed enough to show significant local variations in near-surface earth material types and conditions, in recent changes in vegetation cover, in slope aspect and slope position, and in probable depth to the watertable. Where map scales are too coarse to extract critical terrain and geohazard detail (*e.g.*, tension cracks, areas of disturbed trees, bulged toes of a creeping undercut riverbank slope, fresh-looking scarplets and other slope displacements on active landslides), recent large-scale stereoscopic airphotos are required. All these pipeline terrain and geohazard studies should begin with the interpretation of cost-effective multidiscipline maps.

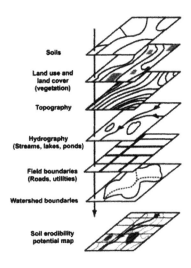

Figure 6 – An example illustrating superposition of different maps and digital data over airphotos or satellite imagery in pipeline route selection and characterization

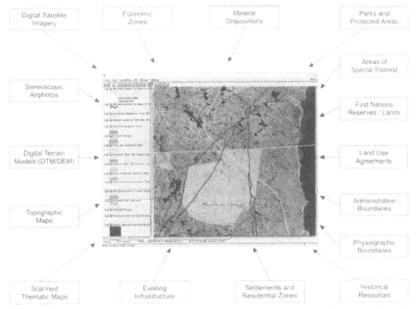

Figure 7 – Example illustrating some of the data inputs involved in selecting and characterizing competing proposed pipeline corridors and right-of-ways

2.3.3 Satellite Imagery Remote Sensing

The first Landsat satellite was launched in July 1972, circling the Earth at an altitude of 920 km in a near-polar, sun-synchronous orbit. Multispectral scanners (MSS) in the first Landsat satellite collected energy reflected from Earth's surface in four wavelength bands with a spatial resolution of 80 m. Three bands are roughly the same as those detected by our eyes, and the fourth waveband range is similar in spectral sensitivity to the three emulsion layers in normal color and color-infrared films. These four bands can be interpreted singly, or combined to create normal color and false-color composite images of Earth's surface. Certain wavebands or combinations of bands are optimum for recording specific targets in the landscape (*e.g.*, soil moisture). Landsat 5 satellite was equipped with a Thematic Mapper sensor (Landsat TM), which collects and records data in seven bands: three bands in the visible range (blue, green and red), three bands in the near- and mid-infrared, all with 30 m spatial resolution, and one band in the emissive (thermal) infrared at 120 m resolution. Because Landsat TM has much better spectral and spatial resolution than Landsat MSS it is more useful and versatile, with applications ranging from assessing changes in soil moisture to discriminating diverse soil and rock types; to mapping vegetation, land use and land cover; to study changes in snow, ice and water cover over time; to detecting and monitoring environmental change. In addition to Landsat imagery, there are many kinds and resolutions of imagery available on the market from commercial satellites. Several satellites provide imagery with a spatial resolution of 1 m or finer.

Because satellite images are digital data, they are readily imported into a geographic information system (GIS) as a georeferenced dataset. In this format, the imagery can be superimposed on other georeferenced satellite images, airphotos and multidiscipline maps for analysis and data presentation. A variety of satellite image processing software and techniques have been developed for quantitative analysis of satellite data.

With an ever-increasing number of Earth-observation satellites in orbit, and increased sophistication in space imaging hardware and image processing, it is possible to acquire up-to-date, specialized, high-resolution satellite images for use in pipeline route selection and characterization studies. The best satellite imagery provides an incredibly detailed synoptic view over large areas. However, information interpreted from this perspective is best correlated and integrated with data interpreted from 3-D stereoscopic airphotos.

2.3.4 Airphoto Remote Sensing

Airphoto terrain analysis relies on the recognition of landforms. The ability to recognize landforms in remotely sensed airphotos is based on the concept that landforms having similar geologic and stress histories and formed under similar conditions of climate, topography, soil and rock material—regardless of geographic location—will have similar airphoto recognition features and similar geoengineering and geoenvironmental properties, characteristics and behaviours for a particular use. Landform feature recognition from airphotos is thus an integral part of an integrated process (Figure 8), which both proceeds and leads to field observation and site investigation of route terrain, and to the impacts that natural hazards and human activities have on pipelines and related infrastructure.

Airphoto identification of terrain types and conditions, including a large variety of minor and major geohazards, is not the same as airphoto interpretation. *Identification* refers to observing, recognizing and naming terrain types and features appearing in airphotos—such as a volcano, a sand dune, a landslide or a river. *Interpretation* of these landforms considers their significance and their suitability and behavior for a particular use or application. For example, an active volcano, an actively migrating sand dune, a slowly creeping landslide, and an actively migrating river channel can all be potentially undesirable terrains and geohazards in pipeline route selection and characterization.

Skilled interpretation of global terrains and geohazards from 3-D airphotos depends on the recognition of landforms and on understanding their genetic geologic processes, environments and stress histories, as well as the soil and rock materials in them. In pipeline route selection studies the airphoto terrain analyst interprets the effects of erosion and deposition by water, wind, ice and gravity agents as well as the effects of harmful climatic change and human actions.

In pipeline route selection studies directed to avoiding natural hazards and improving pipeline integrity, airphoto terrain analysts recognize geoindicators of constructability and maintenance hazards: active riverbank erosion and channel migration, actively sinking ground, a slowly creeping slope. These manifestations of terrain instability may not be shown on maps nor discernible from satellite imagery analysis. Yet they can be interpreted from multidate, multiscale 3-D black-and-white panchromatic aerial photographs. A skilled airphoto terrain analyst detects such information as changes in soil moisture and soil texture from changes in pattern and tone in theoretically 256 shades of grey in black-and-white airphotos.

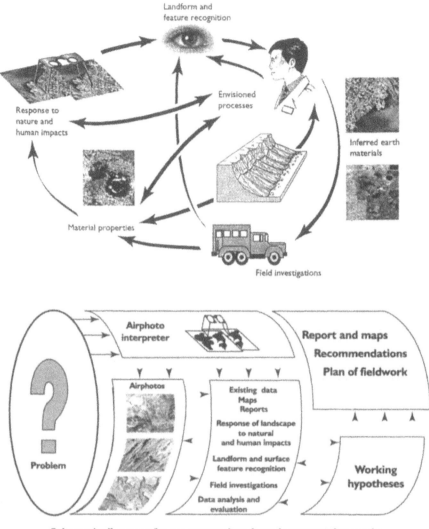

Landform and
feature recognition

Envisioned
processes

Response to
nature and
human impacts

Inferred earth
materials

Material properties

Field investigations

Airphoto
interpreter

Report and maps
Recommendations
Plan of fieldwork

Airphotos

Existing data
Maps
Reports

Response of landscape
to natural
and human impacts

Landform and surface
feature recognition

Field investigations

Data analysis and
evaluation

Problem

Working
hypotheses

Schematic diagram of a representative photo-interpretation study

Figure 8 – Integrating remote sensing, field investigations and pipeline impact assessment

One of the more noteworthy advantages of sequential airphoto terrain analysis is the detection of landscape change over time: change in a creeping landslide or an eroding shoreline, change in the appearance of sediment in water bodies, and human or natural change in land use and land cover. These changes are usually detectable from comparing terrain detail in different ages and scales of airphoto stereopairs—in effect, viewing Earth's landscape features in four dimensions. Integrated spatial and temporal analysis of Earth surface change is possible because, in most areas of the world, airphotos are available in several different scales acquired over the past 60 or more years. Landscape changes can also be detected in multidate, multiscale, multispectral satellite imagery acquired since July 1972, albeit typically at lower spatial resolution than is possible from good quality airphotos.

Because major pipeline route corridors are often long and cross several physiographic and climatic regions, it is customary to assemble strip mosaics from multidiscipline maps, satellite image files and contact airphotos. These strip mosaics, typically about 1 m long, can be placed on a desktop or taped to a wall for viewing a pipeline route corridor and gaining a regional perspective of landscapes along the corridor. In constructing an airphoto strip mosaic, remote sensing analysts use odd- or even-numbered airphotos. The alternate matching stereomate airphotos are then available for 3-D comparison of terrain types, conditions and geohazards on competing pipeline route corridors and along favoured right-of-ways (ROWs) in them.

Figure 9 shows competing alternative pipeline routes selected, and corridors within which the terrain was characterized along the Mackenzie Valley from San Sault Rapids on the north to Fort Simpson on the south, in the Northwest Territories, Canada (Mollard, 1971). Figure 10 is one of the 168 sheets assembled in that study, which includes 14 physiographic and climatic regions from Prudhoe Bay, Alaska, east to the Mackenzie Delta, and south along the Mackenzie Valley to Zama in the northwestern corner of Alberta (see, also, Figure 4). These pipeline route selection and terrain mapping studies illustrate how airphoto-based terrain and geohazard data are combined with site geotechnical and laboratory soil testing data for alternative pipeline route selection and preliminary geotechnical and hydrotechnical characterization. On that study, each airphoto stereogram was shown with corresponding soil and permafrost descriptions and generalized soil profiles along with the ROW geology, topography, vegetation, drainage and geotechnical engineering characteristics. Terrain-mapped sheets similar to Figure 10, and contained in the same large folio, show soil borehole logs above a terrain-typed strip mosaic window with kilometre posts appearing along the ROW and environmental data on fish, birds, mammals, vegetation and archaeology shown in five rows below the airphoto strip mosaic window.

17

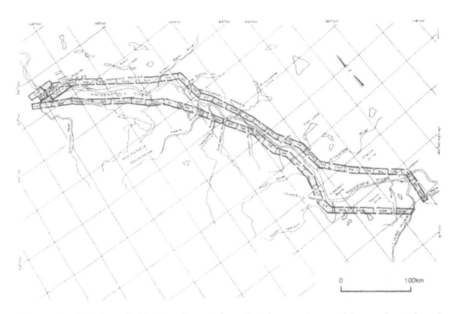

Figure 9 – Mackenzie Valley alternative pipeline route corridors selected and terrain mapped. San Sault Rapids to Fort Simpson and to Liard River, Northwest Territories, Canada (*from* Mollard, 1971)

**GENERAL DESCRIPTION
OF TERRAIN TYPE:** DLc
Thermokarsted lacustrine clay/silt in NW
Outwash deltaic sand/gravel in SE

LANDFORM:
Glacial lacustrine and outwash-delta plains

DEPOSITIONAL ENVIRONMENT:
Lacustrine, glaciofluvial

STRATIGRAPHY (m):
0-2 organics (OL)
3-10 silt/clay (ML-CL) in NW
3-15 sand/gravel (SP-GP) in SE

PERMAFROST

Permafrost zone: discontinuous
Mean annual air temperature: -7 to -4°C
Active layer thickness: 0-3m
Ice-phase description: Vs, Vr, Nbn, Nf
Anticipated excess ice content: low to high
Permafrost table configuration: irregular
Permafrost-affected microrelief features:
thermokarst (thaw basins), beaded drainage

TOPOGRAPHY

Macrorelief (regional): flat, pockmarked
Microrelief (local): peat plateaus
Characteristic slopes: uniform, flat
Characteristic landscape features: thaw basins,
kettle holes, 'fossil' channel scars

VEGETATION

**Natural vegetation zones, formation-types, or
major plant communities:** boreal, mixed
Vegetation structure (layering, life-form):
trees/shrubs/bogs/fens
Tree species composition: conifer/deciduous
Tree stand density: dense / scattered
Common tree height: 3-9m
Evidence of old or recent burns: nil

AIRPHOTO - IDENTIFYING FEATURES

Geomorphic setting: thaw basins, kettle holes
Topographic expression: plain
Tonal characteristics: spotty (NW), dark (SE)
Drainage and hydrographic features: thaw basins
Vegetation characteristics: mixed forest
Present land-use features: nil

TESTHOLE LOGS NUMBERS
3063, 3066, 4320, 57A, 57B

ANTICIPATED ENGINEERING CHARACTERISTICS

Plasticity characteristics: plastic to nonplastic
Organic content: trace to high
Slope stability in thawed condition: unstable in NW,
stable in SE
Settlement potential below pipeline: varies
Trench excavation: caving
Granular borrow: plentiful
Other: good to very poor pipeline conditions

DRAINAGE

**Watercourse types according to size and flow
characteristics:** minor
Channel-bed or land-slope gradients: gentle
Drainage pattern: thermokarst
Drainage density (texture): spotty (NW)
**Sedimentation and erosion character of larger
streams:** nil
Evidence of bank instability along major streams:
nil

INTERPRETIVE COLUMNAR PROFILES

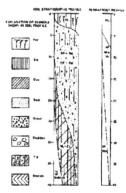

Figure 10 - From one of 168 datasheets summarizing permafrost-affected terrain information along proposed alternative pipeline routes from Prudhoe Bay, Alaska, to northwestern Alberta. This sheet is located near Camsel Bend in the Mackenzie Valley. Total length of pipeline located and terrain mapped: ~3000km (*see* Figures 4and 9) (*modified after* Mollard, 1971).

2.3.5 Ground Truthing of Remotely Sensed Terrain and Geohazard Data on Pipeline Routes

The amount of ground truthing necessary in terrain and geohazard mapping depends on the skill and experience of map, satellite and airphoto interpreters, and the availability of good quality topographic, surficial geology and soil survey maps from which helpful pipeline terrain and geohazard data can be extracted. The amount of ground truthing carried out is also determined by the stage of a pipeline route investigation, among other considerations.

Before the field investigation team begins ground truthing, specific sites to be examined are identified from airphotos, described in considerable detail and discussed to determine the kinds of observations that need to be made. Someone on the ground truthing team should have a good appreciation of the geology of terrain features being investigated, knowing what variations in soil and rock materials can be expected, and their geotechnical properties and characteristics that influence pipeline engineering, construction and maintenance considerations. Some observations made during ground truthing are specifically local or seasonal conditions, such as wet versus dry soils, soft versus firm ground, the likelihood of flooding hazard, surface drainage conditions, frozen versus unfrozen soil material, and the like. In addition, relevant data on landform origin, surficial geology and vegetation cover are also recorded. The airphoto terrain analyst should also be a member of the ground-truthing team.

Initial ground reconnaissance of terrain types and geohazards along a selected pipeline corridor—and often the preferred ROW alignment within that corridor—usually begins with a helicopter flyover, landing at predetermined locations. Where a particular site requiring checking is accessible by vehicle, helicopter aerial reconnaissance may be followed by on-the-ground inspection, traversing significant lengths or all of the ROW on foot.

During on-the-ground reconnaissance, it is desirable to have a copy of the terrain-mapped black-and-white airphoto map at hand. An easy way of recording data for each site visited is to stick a pin through the site location on the airphoto, circle the pinprick hole on the back of the photo and write information about the site with a leader to the circle. Alternatively, the ground truthing team can use an electronic data tablet and GPS unit to record field data. Ground photographs, taken with a camera or video recorder, are usually recorded at the same time.

Ground truthing and site investigation teams should develop a provisional list of significant terrain conditions and geohazards whose features they wish to observe directly before going to the field. Below is a list compiled from past pipeline route selection and characterization studies:

- Landslides (slides, flows, topples and falls) and slope creep
- Wet, soft mineral ground, sinking (subsidence) and settling ground

- Deep soft organic soil deposits, various bog and fen types including ice-rich permafrost in peatlands
- High-swelling (expansive) clay and silty frost-heave soil in permafrost terrains
- Subsurface (piping) erosion
- Migrating river channels and channel sedimentation
- Slide-prone Cretaceous shale (also mudstone and claystone) in actively undercut valleysides and riverbanks
- Hard soil and bedrock requiring blasting
- Highly sensitive liquifiable/collapsible soils, such as marine and glaciomarine silts and clays, and loose calcareous loessial and lacustrine silt and fine sand
- Soluble rock (carbonates and evaporites) features, including solution and collapsing sinkholes
- Corrosive environments in saline soils
- High watertables and fluctuating watertables in fine-grained soils, which can adversely affect the physical, chemical or biochemical behavior of the pipe
- Highly irregular (jagged) bedrock relief, affecting access, travel and blasting costs and rock backfill problems in the trench
- Trench excavation and spoil disposal
- Faults, earthquakes (seismic hazards)

2.3.6 Subsurface Site Investigation

Once ground truthing and selected site observations have been made, additional investigation of subsurface conditions are usually necessary at some sites. Depending on geographic location and the climate, topography and geology of the area in question, specialists in various fields advise the team on the type of exploration equipment to use and on appropriate soil or rock tests to run in a materials laboratory—all part of the geotechnical subsurface investigation.

An experienced pipeline engineering team is necessary to evaluate the negative effects of adverse terrain conditions on pipeline construction, maintenance and operation. Geotechnical engineers, including rock mechanics engineers, are needed to study site soil and rock materials, classify their relevant physical properties, and assess the nature and extent of slope stability hazards. A geophysicist is needed where earthquake or volcanic activity is evident from maps or visible surficial geology features on the ground. A hydrotechnical specialist is required if flooding, erosion, drainage and other natural and human-made hydrotechnical hazards are a concern at creek and river crossings.

Depending on the natural physical and cultural environments under study, site soil and rock investigation may require various geophysical mapping technologies (*e.g.*, electromagnetic methods, ground-penetrating radar), and testhole-boring machines (auger boring, wash boring, rotary coring and cable-tool percussion drilling). The most appropriate testhole drilling methods depend on site conditions and the types of samples required. Because most pipeline trenches extend to depths of 1.5 to 3 m, test pits are commonly used to examine near-surface soil and rock conditions. Soil and rock testing in a laboratory may be required, depending on the situation assessed in the field.

2.4 Nine Global Terrains and Associated Geohazards

The nine different physiographic/climatic terrains discussed in this section are crossed by constructed and proposed oil and gas pipelines in many parts of the world. Significant surface features and ground conditions in these diverse global terrains are largely determined by geology (landforms and surface materials), topographic relief, hydrology and climatic effects. Surface features and controls in turn determine the kinds of geoengineering and geoenvironmental problems that characterize terrain and impact pipeline route selection. Airphoto terrain analysts must be able to recognize a large number of landforms from their airphoto-identifying features, name them and describe their origins, and discuss their significance and behaviour for some particular use or application (Mollard, 1982, Mollard and Janes, 1984, and Mollard, 1995). In this paper, airphoto terrain recognition is required before evaluating the significance of landscapes in pipeline route selection and characterization studies. And in doing this, attention should be paid to construction of the pipeline trench and the installation of the trench backfill, since they can alter natural pre-construction ground conditions.

Essentially all items appearing in the lists of landforms, construction obstacles and geohazards in the nine global terrains can be identified and evaluated for pipeline construction and operational planning to a sufficient degree from interpeting good quality multidiscipline maps, satellite images and black-and-white stereoscopic airphotos. Using these data sources, skilled photointerpreters with relevant engineering and geoscience training and experience provide judgements on avoiding potential problems through appropriate pipeline routing or managing and minimizing potential risks created by identified undesirable (and often avoidable) terrain features and conditions. For many so-called "green field" pipeline development projects, cost estimates in earlier phases of the project are based on these interpretations.

2.4.1 Glaciated Terrains

For over a million years the northern reaches of North America, Europe and Asia have been covered intermittently by a succession of continental ice sheets and local glaciers, which eroded, transported and deposited a variety of soil and rock materials over pre-existing landscapes (Figure 11). In North America, advancing ice sheets eroded parts of Canada's Arctic Islands and much of the North American continental landmass in a broad band south of the Arctic Ocean. Melting glaciers left a legacy of hummocky rolling and ridgy icelaid till in moraines and smaller spreads of waterlaid stratified drift. While some glaciofluvial, glaciolacustrine and glaciomarine deposits were laid down in direct contact with the ice, most of this waterlaid material was carried and deposited far from the retreating ice margin. Today, glaciers cover about 15 million square kilometres of the Earth's surface, mainly in the Antarctic, Greenland, Alaska, and Canada's Arctic Islands.

Not only in northern North America, but also in northern Europe and Asia, landscapes created by continental ice sheets and their products are characterized by numerous lakes, bedrock exposures, and different varieties of supraglacial, englacial and subglacial till and ice-contact and proglacial stratified deposits. The northern parts of these continents also have extensive permafrost. South of Precambrian Shield terrains, where discontinuous silty sandy gravelly till and stratified drift overlie eroded hard rock types, the glaciated landscapes are dominated and characterized by sandy, silty and clayey till deposits that overlie extensively flat-lying younger and softer Paleozoic and Mesozic sedimentary rocks. The grain-size compositions of these tills reflect the compositions of their source bedrock, which was eroded by advancing glacial ice. In general, glacier-eroded clay shales give rise to clayey till, carbonate rocks to silty carbonate-rich till, sandstones to sandy till, and igneous and metamorphic rocks to a silty sandy gravelly till with little clay—each till type containing glacial boulders, a potential problem in ditch trench excavation in some locations.

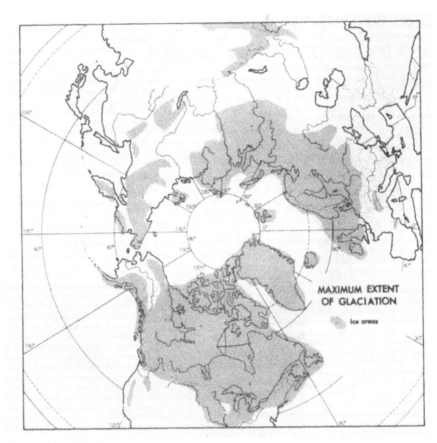

Figure 11 – Maximum extent of glaciated terrain in the northern hemisphere, including present ice areas (*from* Hoyt, 1967)

Stratified sand and gravel deposits occur in ridgy and hummocky ice-contact glaciofluvial terrains (kames and eskers) and in level and kettled proglacial glaciofluvial terrains (outwash plains and outwash deltas). All are potential sources of construction aggregate and groundwater supplies for pipeline construction and maintenance. Extensive level and undulating fine-grained (silty and clayey) glaciolacustrine and glaciomarine sediments can present serious slope instability problems along pipeline routes, especially where they have been deeply channeled by catastrophic glacial meltwater flows, forming small and large spillway valleys that served as drainageways for escaping glacial meltwater. Most of these meltwater valleys are partly filled with stratified fine and coarse fluvial and lacustrine waterlaid sediment. Small "misfit" streams in them migrate over wide floodplains, their channels in silty and sandy deposits shifting and eroding their streambanks. In addition, silty sand and fine sandy riverbeds are particularly prone to downward riverbed erosion during flood conditions while river banks are eroded sideways. Where meltwater channels have eroded through a cover of glacial deposits and into the underlying Cretaceous and Tertiary bentonitic clay shale and mudstone, large rotational slump and translational slides are common occurrences along valley sides. These landslides commonly creep at rates of a few centimetres to as much as 1 m a year, especially where the valley sides and riverbanks are actively eroded by rivers.

The following list of glaciated terrain landforms, obstacles and geohazards affecting pipeline route selection can be identified and evaluated from aerial and satellite imagery. While some features are minor obstacles to pipeline construction and operation, others are major and significantly affect pipeline integrity (*see, also,* the accompanying list of additional reading references). Potential geohazard sites along the selected pipeline route are investigated to decide whether to move the route or correct, manage or minimize any integrity concerns.

- Rough, bumpy rock-controlled relief, resulting in additional heavy rock blasting and excavation for the pipe trench. This is a common terrain in Precambrian Shield and glaciated mountain terrains.

- Moderate relief hummocky moraine, with locally steep-sided mounds and ridges, and kettleholes commonly that flood during spring snowmelt

- Moderate relief end moraine ridges

- Ice-thrust moraine, which often has rafted inclusions of dislocated bentonitic marine shale that can be unstable and prone to slide activity

- Deeply kettled kame moraine, where spring runoff collects in kettleholes leading to seasonally fluctuating watertables

- High relief drumlins

- Rögen (ribbed) and de Geer (washboard) moraines, where surface boulders 1 to 2 m in diameter create obstacles to pipeline construction

- Permafrost-affected glaciated terrains having potential for frost heave and thaw settlement (*see* Section 2.4.3)

- Highly compressible peatland terrain concentrated in Canada's boreal forest (*see* Section 2.4.4)

- Soft, compressible settlement-prone lacustrine and marine clay

- Highly expansive glaciolacustrine clay that causes pipeline heave during seasonal and cyclical wet and dry periods

- Large landslides along river valleys, including actively creeping slopes at undercut drift-covered Tertiary and Cretaceous bentonitic shale within and beyond glaciated terrains (Figures 12 and 13)

- Large lateral-spread failures in metastable sensitive marine and glaciomarine silt and clay deposits, where excessive rainfall and snowmelt infiltration, toe erosion and earthquake and other ground shaking can trigger sudden diastrous slope movements
- Flooding along abandoned glacial spillways, other glacial meltwater channels, tunnel channels and postglacial river valleys
- High capillary rise in marine and lacustrine silt, leading to corrosion in saline soils
- Playas and saline soils that represent a corrosion hazard
- Ongoing glacial rebound from ice-sheet melting and crustal unloading

Figure 12 – Small-scale airphoto showing two pipeline routes (fine white arrows), which avoid a large actively moving landslide in till overlying slide-prone Cretaceious clay shale with thin bentonite seams. South Saskatchewan River near Outlook, Saskatchewan
(see **an enlarged view of this landslide in Figure 13)**

26

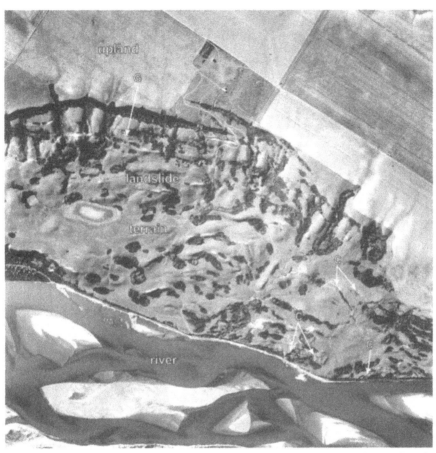

Figure 13 – Large-scale airphoto view of retrogressive slumping in glaciated terrain (*see* Figure 12). This slide was avoided in pipeline route selection. South Saskatchewan river, south of Outlook, Saskatchewan, Canada (C: fresh cracks; G: sinking graben below headscarp)

2.4.2 Fluvial Terrain

The term "fluvial terrain" includes landscapes that have been eroded by running water or that have been built up by accreting deposits over time. Both impact pipeline location and design (*see* Chapter 3 titled "Open Cut Water Crossings"). Common river channel types can be identified and their behaviour provisionally predicted and evaluated from controlling factors interpreted from map and 3-D airphoto terrain analysis (Figure 14). It is also important to have a basic understanding of typical environments and typical bed and bank materials that create different planforms and types of river channels (Figure 15), which affect channel bank and riverbed stability tendencies. These kinds of data are helpful in selecting pipeline routes and avoiding pipeline integrity problems at valley and channel crossings.

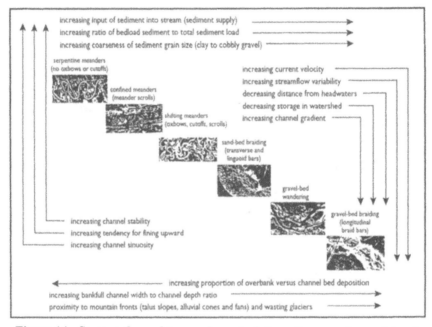

Figure 14 - Stream channel types, characteristic patterns and a variety of factors influencing channel type, planform pattern and behaviour tendency (*modified from* Mollard, 1973)

28

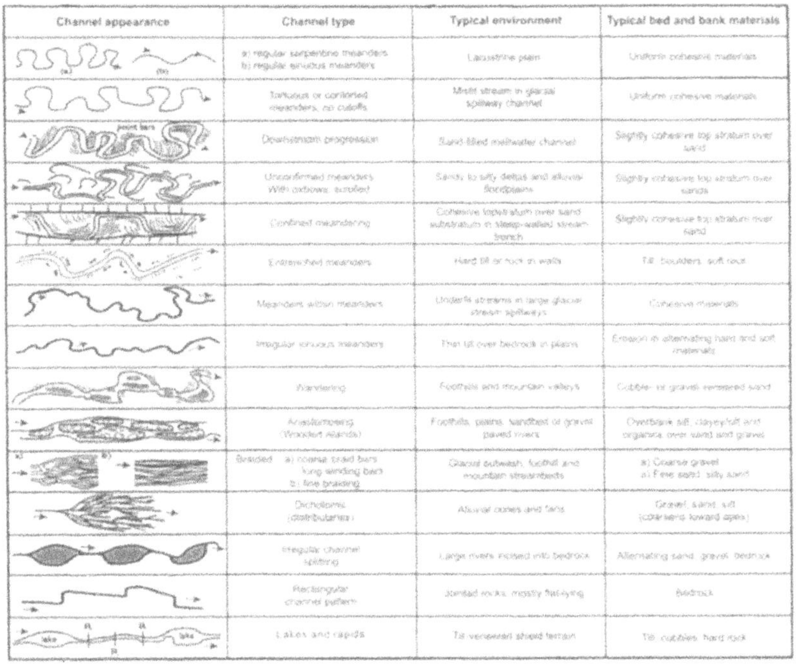

Channel appearance	Channel type	Typical environment	Typical bed and bank materials
	a) regular serpentine meanders b) regular sinuous meanders	Lacustrine plain	Uniform cohesive materials
	Tortuous or confined meanders, no cutoffs	Misfit stream in glacial spillway channel	Uniform cohesive materials
	Downstream progression	Sand-filled meltwater channel	Slightly cohesive top stratum over sand
	Unconfined meanders with oxbows, scrolled	Sandy to silty deltas and alluvial floodplains	Slightly cohesive top stratum over sands
	Confined meandering	Cohesive topstratum over sand substratum in steep-walled stream trench	Slightly cohesive top stratum over sand
	Entrenched meanders	Hard till or rock in walls	Till, boulders, soft rock
	Meanders within meanders	Underfit streams in large glacial stream spillways	Cohesive materials
	Irregular sinuous meanders	Thin till over bedrock in places	Erosion in alternating hard and soft materials
	Wandering	Foothills and mountain valleys	Cobble- or gravel-rendered sand
	Anastomosing (Wooded islands)	Foothills, plains, sandbed or gravel paved rivers	Overbank silt, clayey silt and organics over sand and gravel
	Braided a) coarse braid bars long winding bars b) fine braiding	Glacial outwash, foothill and mountain streambeds	a) Coarse gravel a) Fine sand, silty sand
	Dichotomic (distributaries)	Alluvial dunes and fans	Gravel, sand, silt (coarsens toward apex)
	Irregular channel splitting	Large rivers incised into bedrock	Alternating sand, gravel, bedrock
	Rectangular channel pattern	Jointed rocks, mostly flat-lying	Bedrock
	Lakes and rapids	Till veneered shield terrain	Till, cobbles, hard rock

Figure 15 - Stream channel types, environments, and typical bed and bank material

Some fluvial channels occupy narrow valleys and have little or no floodplain while others actively migrate back and forth across wide floodplains, forming meander scrolls and oxbows, eroding and undermining pipelines. The airphoto terrain analyst must not only recognize different channels but also recognize a variety of drainage patterns (Figure 16), created by the erosion and deposition of different kinds of soil and rock material in different geologic, topographic and climatic environments. Steep stream channels in mountain and foothill landscapes erode coarse cobbly and gravelly sediment from alluvial fans and talus deposits, and carry the eroded sediment downstream where it is deposited in wide, shallow braided channels with erodible banks and shifting gravelly riverbeds. Similar channels develop downstream from retreating mountain valley glaciers.

Stream channels that form in deep valleys with narrow floodplains are usually characterized by steep gradients, bedrock waterfalls, cascades and rapids. These fluvial features contrast with those found along wide abandoned glacial spillway valleys, formed by large flows of glacial meltwater whose characterizing features include slowly aggrading overbank floodplains, migrating meanders, pointbars,

29

meander scrolls, oxbow lakes and cutoffs, and backswamps (Figure 17) as well as natural levees, crevasse splays, marshes and perched floodbasins. Other fluvial landforms that can be significant in selecting and characterizing pipeline routes are actively eroding or depositing alluvial fans, cones, terraces, deltas and pediments.

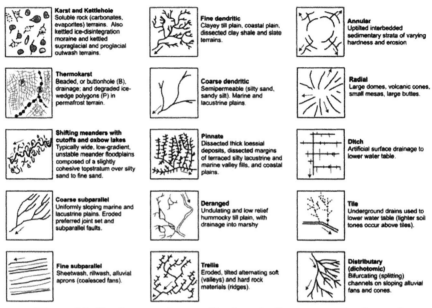

Figure 16 - Drainage patterns affecting pipeline route selection and characterization

Pipeline river crossings in different valley and channel types represent a major challenge to hydrotechnical engineers, who determine whether the pipeline will be stable after being installed in alluvium under a riverbed, or whether a tunnel must be bored through bedrock underlying the river channel and its fill of fluvial and lacustrine sediment (*see* Chapter 3 titled "Open Cut Water Crossings").

Straight channel sections are favored over channel bends at pipeline crossing sites. Landslides encroaching on valleysides and riverbanks that are being actively undercut can be critical pipeline stability locations the terrain analyst identifies, evaluates and avoids to improve pipeline integrity wherever practical.

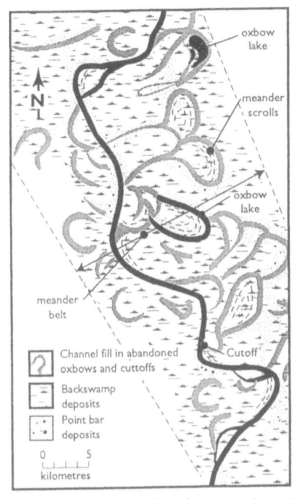

Figure 17 - Features occurring on a shifting, freely meandering low-gradient stream eroding dominantly cohesionless (*e.g.*, silty fine sand) deposits in a wide floodplain (Assiniboine River valley, southwestern Manitoba)

The following fluvial features are routinely identified from remotely sensed maps, aerial and satellite imagery. Wherever possible, pipeline routing should avoid:

- Highly mobile riverbed sediments
- Migrating channels, especially outer banks at meander bends
- Shifting meanders characterized by many cutoffs, oxbows and scroll bars, usually indicating a highly unstable channel

31

- Dormant and active slope failures along riverbanks and valleysides
- Actively undercut valleysides and riverbanks, a major cause of slope creep in landslide-prone environments
- Alternating hard and soft tilted layers of sedimentary bedrock at river valley crossings
- Thick accumulations of loose riverbed silt and fine sand subject to deep scour during major flood flows
- Channel diversions that detrimentally affect stream erosion and deposition
- Human alteration of river channels, such as channel straightening, dredging, and reservoir construction and degradation of sand riverbeds below dams
- Geomorphic and hydrologic controls on avulsions, causing flooding and scour
- Landscape fluvial dissection: high density of drainage courses, gullying
- Erosion at narrow constrictions on floodplains
- Undesirable effects of channel shifting on actively eroding and depositing alluvial fans
- Silted-up, abandoned distributary channels on birdfoot deltas
- Loose collapsible silt and fine sandy soils beneath alluvial floodplains
- Erodible bed and bank materials, consisting of cohesionless silt and fine sand alluvial deposits
- Channel scour from a wide variety of causes (Figure 18)
- Local riverbed aggradation, causing overbank flooding
- Anomalous variation in river channel width from various causes
- Excessive ongoing (active) fluvial dissection of cohesive materials
- Large boulders in channel beds
- Rapids and falls
- Locations where ice dams, jams and icings may form in northern rivers

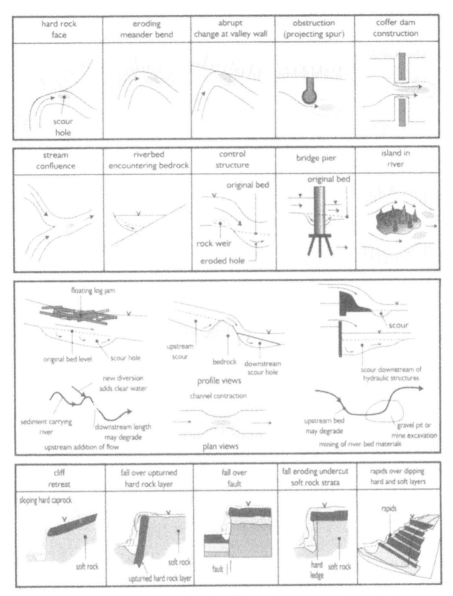

Figure 18 - Plan and profile sketches illustrating causes of riverbed scour (top and center groups of river-scour are sketches are from C.R. Neill, *pers. comm.***)**

33

2.4.3 Permafrost-affected Terrain

Permafrost is defined as the thermal condition in soil or rock where the temperature persists below 0°C over at least two consecutive winters and the intervening summer. In North America, northern Europe and northern Asia, permafrost becomes increasingly widespread northward toward the Arctic Ocean (Figure 19). In the Canadian arctic islands, permafrost reaches depths exceeding 800 m.

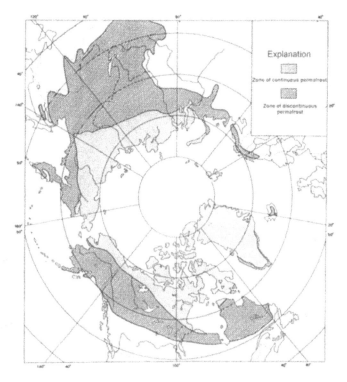

Figure 19 - Map of permafrost distribution in the northern hemisphere
(*from* Péwé, 1975)

In North America, researchers have separated permafrost-affected terrain into four zones. The continuous permafrost zone exists everywhere beneath the land surface except under deep lakes and in newly deposited unconsolidated soil materials. Widespread discontinuous and scattered discontinuous zones are found where permafrost exists along with unfrozen ground, and where unfrozen layers occur within the permafrost itself. In the discontinuous zone, permafrost is commonly found on shaded north-facing slopes and in raised Sphagnum-rich peat plateau bogs with small thaw-collapse basins (Figure 20).

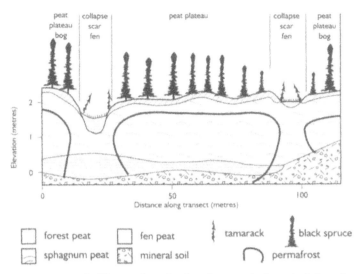

☐ forest peat	☐ fen peat	↓ tamarack	↑ black spruce
▨ sphagnum peat	▨ mineral soil	∩ permafrost

Figure 20 - Schematic illustrating the development of peat plateau bogs and thaw-collapse scar fens, a common peatland terrain in Canada's boreal forest

In permafrost-affected terrain, the active layer (which freezes and thaws annually between ground surface and top of the permafrost table) varies from less than 1 m to about 10 m in depth locally. The active layer depth can also vary from year to year. It overlies permafrost (frozen ground), which may have bodies of unfrozen ground called taliks (Figure 21). Because the permafrost table is usually close to ground surface in continuous permafrost, reducing the ground permeability to zero, rainfall and snowmelt runoff is forced to flow over the ground surface and through a relatively thin active layer below.

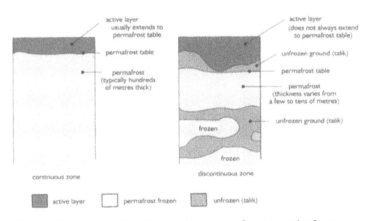

Figure 21 - Schematics illustrating permafrost terminology

35

Ice jams, icings in streams, creep and caving of ice-rich streambanks are all concerns in pipeline route selection, as are ice-rich and ice-cored mounds and terrain environments that create frost heave, thaw settlement and thermal erosion.

Thaw depressions (thermokarst) create a landscape characterized by spotty subequally spaced ponds and drained basins of roughly similar size. The ponds and basins commonly create a speckled-looking landscape that is easily recognized in airphotos. Thermokarst features are a common occurrence on permafrost-affected fine-grained glaciolacustrine, lacustrine, glaciomarine and marine deposits. South of the continuous zone in the widespread discontinuous zone and in the southern fringe of permafrost distribution, peat plateau bogs with thaw-collapse basins (Figure 22) and with ice-wedge polygons (polygonal peat plateau bogs) are diagnostic features of permafrost-affected landscapes in airphotos.

The harmful effects of climate change on the stability of pipelines can be a serious problem in permafrost-affected terrains. These effects vary with geographic location and with terrain type and conditions. The different terrains have been comparatively rated from stereoscopic airphoto interpretation in selecting and characterizing competing pipeline routes crossing continuous and discontinuous permafrost zones in northern Canada.

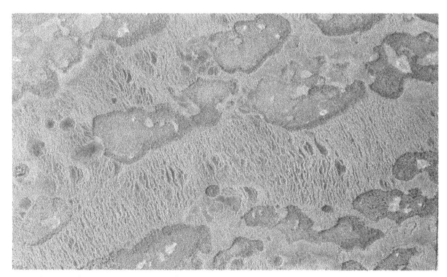

Figure 22 - Airphoto showing slightly elevated (0.5 to 1.5 m) treed peat plateau bogs and surrounding string (ribbed) fens. The raised bogs have small whitish thaw-collapse scars, are frozen and contain ground ice (permafrost). The fens are unfrozen and have a watertable at or near ground surface

For purposes of pipeline route selection studies, permafrost terrain features may be characterized as either favorable or undesirable. The following permafrost terrain features including obstacles and hazards (all of which could be characterized as unfavorable) are recognizable in good quality 3-D airphotos, and should be evaluated to determine whether they can be circumvented where practical:

- Low-centre and high-centre ice wedge polygons
- Various sorted and nonsorted patterned ground stripes, steps, circles, nets, polygons and hummocks, indicating surficial materials subject to intensive frost action processes
- Different origins and sizes of ice-rich/ice-cored mounds, including open and closed pingos, palsas, earth hummocks, frost blisters, frost boils, peat plateau bogs and polygonal peat plateau bogs, which heave when ice in them is aggraded and settle when the ice in them is thawed
- Felsenmeer and blockfields of fractured bedrock that can be difficult to clear from the right-of-way
- Bimodal flows, where thawing horseshoe-shaped backscarps are nearly vertical and the resulting mud slurry flows out of the gently sloping flow bowl
- Multiple retrogressive thaw slumps
- Active layer detachment (skin) flows
- Falls and topples
- Solifluction creep features (sheets, terraces, lobes)
- Creeping ice-rich streambanks and valleysides
- Creeping rock glaciers and talus slopes
- Frozen/unfrozen thermal transitions in discontinuous permafrost
- Oriented lakes, some partly and others entirely sediment-infilled
- Beaded (buttonhole) streams, roundish thaw holes at ice-wedge intersections along small streams
- Horsetail drainage (closely spaced, subparallel surface drainage runoff lines)
- "Drunken forest" of leaning trees caused by thawing of permafrost in ice-rich peat bogs and by undermining lake, pond and stream shore cliffs (Figure 23)

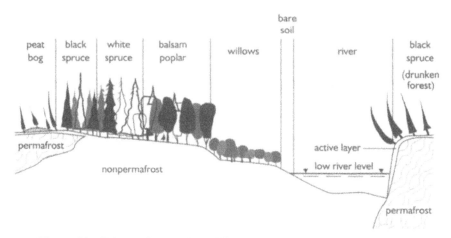

Figure 23 - Schematic showing different vegetation types common in widespread discontinuous permafrost areas

2.4.4 Peatland / Wetland Terrains

Peatlands (bogs and fens) are one of the more common and important wetland landforms, which also include swamps and marshes. All can influence pipeline selection and characterization. Canada has the most extensive area of peatland in the world, covering roughly 12 percent of the country's land surface: about 110 million hectares, or 420,000 square miles. Moreover, Canada is a country with many existing and proposed major pipelines located in high watertable, high compressibility peatlands with contiguous frozen and unfrozen peat landforms, the majority of which are located in the boreal forest. The Scandinavian countries and parts of the United Kingdom and Russia (Siberia) also have large areas and different types of bogs and fens.

A cool climate is an important factor controlling peat development, especially where low evapotranspiration allows rainfall and snowmelt runoff to accumulate in waterlogged basins and other depressions underlain by fine-grained waterlaid mineral soils having low permeabilities and thus impeded drainage. Peat landscapes rarely develop on granular landforms, or on sloping ground where surface runoff is rapid. Exceptions are maritime coastal areas having high rainfall, extensive fog and low evaporation. In this environment, peat can even form on relatively steep slopes and on well-drained coarse mineral soils.

Peatland is a waterlogged, highly compressible landform having a compression index of about 1 per cent of the peat's natural water content, which can reach 1000 per cent and more. Peat depths reach as much as 10 m and more. Although unfrozen peatland is typically saturated to or near ground surface over the summer months, watertables can drop dramatically during long hot seasons and protracted

38

droughts. Where peatland forms small isolated islands surrounded by mineral soil, they can often be circumvented through lateral realignment during pipeline routing. In northern areas where extensive peatlands can't be avoided in pipeline routing, pipeline construction is usually undertaken in winter when construction equipment can travel on frozen peat surfaces. Where peatlands form extensive bog blankets mantling mineral-soil surfaces, they are frequently used as "winter roads" when the ground is frozen. Ice roads over alternately frozen bogs and unfrozen fens are constructed by flooding packed snow to give a smooth ice riding surface. Watertrack fens usually freeze up a month or so later than bogs, which shortens the length of winter road trafficability.

A useful geoindicator of shallow versus deep peat is the envisioned topographic shape of the mineral soil surface underlying peatland. Shallow saucer-shaped mineral surfaces underlying peat commonly correlate with thinner peat depths. Sharply depressed bowl-shaped hollows in the underlying mineral sediment surface correlate with thicker peat, especially in bogs (*e.g.*, over deep kettleholes in outwash deposits and deep glacier-scoured rock basins). In general, bog peat landforms are significantly deeper than fen peat landforms. Bogs are also a characteristic feature in northern coastal areas in North America, Europe and Asia.

Peatland researchers have classified roughly 20 types of bog and 20 types of fen. Many form bog-fen or fen-bog complexes, depending on whether the bog or the fen component dominates. In many cases, peatland is unavoidable in pipeline routing, and it is necessary to choose the shortest or the least hazardous route through it. The following simplified peatland classification has been used to remotely sense, map and evaluate extensive peatland types in the James Bay Lowland of northern Ontario.

- Forested, semi-forested and nonforested black spruce/Sphagnum-dominated bogs, typically with slightly raised (domed and plateau) surfaces (Figure 24)

- Watertrack and ribbed (string/ladderlike) sedge-tamarack dominated fens, typically occurring on extensive smooth, low-gradient slopes (horizontal fens) and in channelways (channel and watertrack fens)

- Sweeping watertrack fens enclosing teardrop-shaped and darker-toned forested bog islands, typically on gentle slopes

- Spring fens, where groundwater breaks through to ground surface giving a finely speckled pattern of small ponds and small raised peat forms

- Bog-fen and fen-bog complexes

- Peat plateau bogs with ground ice (permafrost), speckled with small thaw basins called "collapse scars" from the meltout of ice

- Polygonal peat plateau bogs with ice-wedge polygons
- Palsas (ice-cored peat mounds)
- Northern swamps

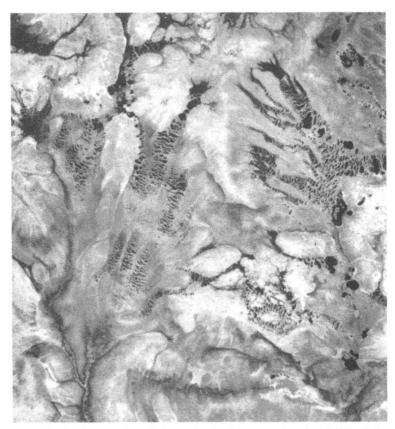

**Figure 24 - Light-toned oval-shaped raised bogs and string (ribbed) fens.
Northern Ontario, Canada
(National Air Photo Library airphoto A14081-108)**

2.4.5 Mountain Terrains

Typically, mountain ranges with intervening valleys, lakes and smaller basins are located along continental margins where high rainfall and snowfall and a generally wet climate predispose the high relief and steep slopes to sudden slope failures, some small and others huge. Dormant slope failures can be easily activated (and reactivated) by natural and human activities, creating pipeline integrity and public safety hazards. Examples can be found in the West Coast ranges of North, Central and South America, the Guiana Highlands of South America, the Alps, the Urals, the Caucasus, the Himalayas, Hindu Kush, Tian Shan, and parts of Indonesia, among others in the Americas, Europe and Asia.

Figure 25 lists a variety of mountain terrain hazards along a short stretch of the proposed Alaska pipeline route, at Sheep Mountain where the Slims River delta enters Kluane Lake, Yukon, Canada. Three major slope failures on Sheep Mountain above the Alaska Highway and the proposed Alaska pipeline route are a massive rock avalanche, a rock glacier and a debris torrent, which failed a recorded eight times in 1,200 years, forming a debris cone above the highway (Figure 26).

AIRPHOTO-IDENTIFIED AND MAPPED TERRAIN GEOHAZARDS
ADJOINING THE ALASKA HIGHWAY AT SHEEP MOUNTAIN,
YUKON TERRITORY, CANADA

- Large rockfall avalanche
- Rock slump
- Rockslide
- Rock glacier
- Debris flow and debris torrent
- Slims River delta-front sediment liquefaction
- Snow avalanche
- Blocky rubble hurtling down the mountainside
- Jokulhlaup (ice-dammed lake outburst)
- Flooding
- Torrential rain
- Earthquake
- Dust storm
- Compound geohazard

Figure 25 - Mulitple geohazards assessed from airphotos and ground reconnaissance along the proposed Alaska Highway pipeline route. Sheep Mountain, Kluane Lake, Yukon Territory, Canada

The following partial list of pipeline hazards in mountainous terrains, which includes the effects of slope-instability processes and environments inferrable from maps and remotely sensed imagery, should be assessed and avoided where possible in pipeline routing.

- High slopes, steep gradients, often with sharp breaks in slope including sidelong (steep sidehill) ground

- Glacier-ice eroded "oversteeped" mountain and valley slopes, hanging valleys and fiords

41

- Neotectonic uplift (*e.g.*, glacier-melt unloading and glacial rebound) and subsidence, mostly small and variable in magnitude along coastal mountains and other ranges, yet significant enough in some locations to predispose the terrain to landslide activity
- Rock structures such as folds, faults and fractures, especially fault-rupture displacements in deformed layered rocks, caused by strong-motion earthquakes
- Sackungen lineaments (upslope-facing scarps) on upper mountainsides, which signify slope dislocation and local topographic barriers
- Various earthquake-associated topographic features, *e.g.*, linear fault scarps, including offset topographic and drainage features
- High rainfall areas in coastal regions
- Periodically flooded mountain valley bottoms
- Permafrost-affected slopes in high latitude and high altitude mountains
- Rock avalanches (Figure 26)
- Rock glaciers (Figure 26)
- Metastable debris flows and torrents (Figure 26)
- Rock slides and rock slumps
- Metastable mudslides and mudflows
- Metastable topples and rock falls
- Snow, slush and debris avalanche tracks
- Flows variously referred to as earthflows, lateral spreads, spreads, flowslides, slab failures, piston failures and liquefaction failures
- Recurring ravels and small slips on actively shedding undercut steep slopes, which may suggest future mass movements
- Lahars on active volcanoes
- Solifluction creep in high mountain altitudes and high latitudes
- Forest fires below the timberline that destroy trees and their slope-stabilizing root systems

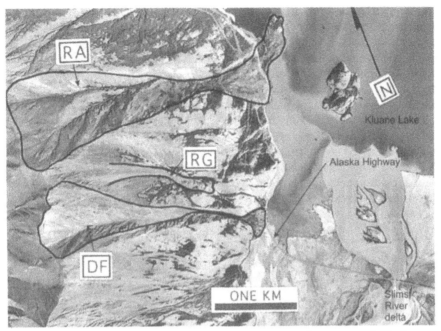

Figure 26 - Major slope movement features above the proposed Alaska Highway pipeline route at Sheep Mountain, Kluane National Park, Yukon, Canada. Legend: RA, rock debris avalanche track; RG, rock glacier track; DF, debris flow track. Note the large failure material source areas on the upper mountain side and the downslope debris spoil. The DF slope is reported to have failed 8 times in the past 1,200 years

2.4.6 Volcanic Terrains

While active volcanic terrains are obviously poor places to locate a pipeline, even long extinct volcanoes pose their own hazards. Some 80 per cent of the world's active volcanic centres are located on or near the Pacific Rim. Known as the "Ring of Fire," this area extends along the west coast of the Americas from Chile to Alaska and the Aleutian Islands, and along the east coast of Asia south to the East Indies and New Zealand. About 14 per cent of the world's active volcanoes are located in the Indonesian Archipelago. Scattered volcanoes also occur along the Mediterranean Sea's north shore.

Volcanic terrains consist of erupted airfall fragmental debris and lava flows. Volcanic materials ejected into the air, ranging in size from dust to rounded and angular blocks larger than 64 mm across, are called pyroclastic and tephra deposits. Some volcanic terrains also consist of interbedded pyroclastics and lavas. Lava flows erupting through systems of deep fissures spread outward in sheets, piling up to a thousand metres or so in thickness, forming extensive flood basalt plateaus and tablelands. Sheet-upon-sheet of stacked basaltic lava outpourings are exposed to view on near-vertical canyon cliffs in the states of Washington, Oregon and Idaho in the northwestern United States (Figure 27). As the molten lava congealed into hard basaltic rock, it developed hexagonal columns that resemble piles of cordwood set vertically on end (Figure 28). Locations where canyons have tall cliffs, ubiquitous high watertables and highly plastic residual clay surfaces on buried subaerially weathered lava beds are susceptible to huge landslides. Networks of open fractures (fissures) in the flood basalts are capable of storing large volumes of groundwater.

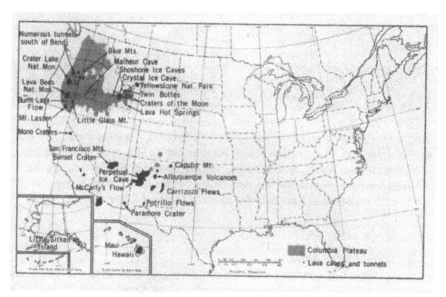

Figure 27 - Volcanic terrains in the northwestern United States, Arizona and New Mexico (*from* Snead, 1972)

Below is a checklist of terrain obstacles and geohazards that remote sensing terrain analysts can expect to find, and should avoid as much as possible, when locating and planning pipeline routes over volcanic terrains.

- Volcanic cones with steep slopes (commonly 30° to 35°), which can become unstable
- Potential volcanic dome and cone lava eruptions

44

- Pyroclastic fragmented debris that weather to plastic clay and erode into intricately dissected terrains, which can be difficult and costly to cross
- Former and possible future outpouring of hot lava flows interbedded and mixed with fragmental airfall ejecta. Such locations are hazardous where volcanoes are overlain by thick snow-covered steep slopes
- Wrinkled, ropy lava tongues, especially young lavas and lava fields located in dry regions, influencing the volume of rock blasting required in trench digging
- Lava "squeeze-ups" and pressure ridges that can create a bumpy land surface
- Lava channels, tubes and tunnels that may collapse
- Loosely cemented ashfall deposits that are subject to deep gully dissection on steep slopes
- Young weakly cemented pyroclastic deposits that are easily and excessively eroded
- Locations near active volcanoes associated with high concentrations of volcanic ash and dust
- Toxic steam and gases emitted from active volcanoes (*e.g.*, Mount St. Helens), which are corrosive and a potential health hazard to humans and other living things
- Locations with a high seismicity hazard (strong-motion earthquakes)
- Locations where there is a prospect of significant fault displacement movements during large magnitude earthquakes
- High-volume rock excavations, requiring blasting in the digging of pipeline trenches
- Debris avalanches on cinder cones
- Lahars consisting of hot lava flows and pyroclastic ejecta on cones
- Near-vertical stairstepped, sawtoothed cliffs on canyon walls
- Landslides on upper basalt cliffs, creating piles of loose broken columns of talus below
- Topples and tumbling of talus blocks (broken basalt columns)
- Highly plastic residual clay between lava flows that induce landslide activity
- Large groundwater springs
- Bumpy relief over sharp depressions and lava mounds, increasing the amount of rock blasting

**Figure 28 - Close-up and distant views of columnar basalt,
Washington State, USA**

2.4.7 Coastal Terrains

Water currents tend to be more important than waves in creating erosion and deposition in rivers, whereas waves are more important than currents in creating erosion and deposition along ocean shore zones, developing different coastal types and continuously modifying their geometry in plan and profile (Figure 29).

The effects of wave action on shore zones in coastal areas are strongest between about 10 m below and 10 m above mean sea level. Beyond about 10 m below mean sea level, the wave base depth is insufficient to move sediment on the ocean floor. Along shoal coasts, waves break far from the shoreline in a low wave-energy environment, where shore and ocean bottom erosion are lower.

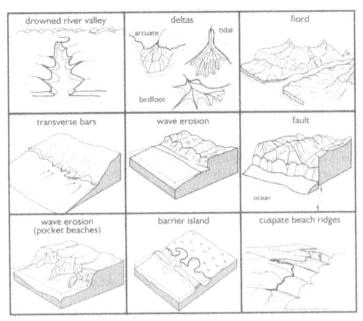

Figure 29 - Cartoon sketches illustrating coastal environments associated with nine shoreline types

Although they are relentlessly ongoing, the shorezone-altering effects of ocean wave and current erosion and deposition usually don't exceed more than about 10 m above mean sea level. Exceptions are locations where tsunamis and wave action erode the toe of high shoreline bluffs, resulting in bluff failure, removal of failure spoil, and landward retreat of the bluff from ongoing erosion. Along ocean and lake shores, waves bend toward and erode headlands and other shoreline promontories. Eroded coarser sediment is transported into adjoining inlets and sheltered bays by wave-generated alongshore currents. In consequence, pocket beaches and sand dunes are characteristic depositional features in small coastal inlets. In contrast, barrier spits, islands, passes, flood and ebb tidal deltas, washover fans, lagoons and tidal marshes, commonly with associated sandy beaches and dunes, are characteristic features of large coastal embayments. Here, longshore currents strike gently shelving shores at an angle, moving sand particles along the beach while high velocity rip currents erode channels in the nearshore ocean floor, driving sediment directly offshore.

Onshore and nearshore underwater slopes on cones, fans, deltas, beaches, spits, bars and barrier islands require the special attention of airphoto terrain analysts because they are continually modified by ongoing shore erosion and deposition processes. To estimate the average rate of annual shore recession or advance, changes in the shoreline position are plotted and measured from airphotos and high-resolution satellite images acquired at the same location a few years to several decades apart.

Deeply buried offshore clean sand shorelines along some of the world's largest estuarine, birdfoot, tidal and arcuate deltas (Figure 30) are sites of major oil and gas exploration and discovery, and therefore the sites of subaerial onshore and subaqueous nearshore and offshore pipelines. Examples are the Mackenzie, Mississippi, Tigris/Euphrates Shatt-al Arab, Ganges, Ob, Niger, Congo-Luando and Amazon deltas. As major oil and gas fields are explored and developed in these coastal environments, pipeline route location and planning can be a major challenge. These coastal environments are particularly exposed to the ravages of tsunamis, hurricanes, strong winds, high waves and floating ice, any of which can create enormous pipeline design, construction and operation problems.

Coastal features to look for and assess from maps, satellite and airphoto imagery include:

- Evidence of past tsunami damage
- Evidence of hurricanes, surging tides and wind damage (for example, the Mississippi Delta)
- The extent of coastal erosion and deposition change over time
- Permafrost conditions, both onshore and offshore in the Arctic Ocean
- Landslides and creep movements along eroding shore bluffs
- Subaqueous flowslides, often triggered by seismic activity
- Topples and falls along eroding cliffs
- Shifting (eroding and depositing) distributary beds on shorezone fans and deltas
- Shoreline recession and progradation, assessed from sequential airphoto measurements
- Heavy flooding of onshore large fluvial-deltaic plains
- High salinity and related pipeline corrosion
- Lagoons, tidal marshes, swamps and other wetlands having thick organic deposits
- Soft, weak compressible onshore and offshore silty and clayey deposits, requiring close balancing of pipeline weights
- Expansive clays causing pipeline heave
- Complex system of fault and fracture networks

48

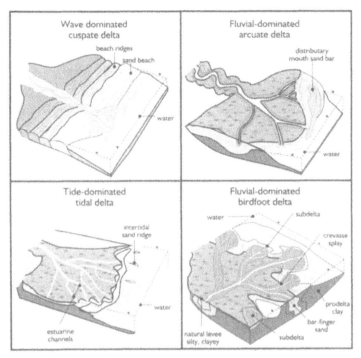

Figure 30 - Recognition features on four main coastal delta types and depositional environments

2.4.8 Karst Terrains

The term "karst" is used to characterize sinkhole topography produced by surface and subsurface water flow, dissolution, subsidence and underground cave and cavern collapse in carbonate and evaporite strata. While dissolution action is mainly chemical, caused by flowing acidic (CO_2) groundwater, it is also partly mechanical, the effects of flowing water physically wearing away carbonate rock. Although gypsum, anhydrite and halite evaporite formations are much more soluble than are carbonate (limestone and dolostone) formations, the more limited exposure of evaporites (except for very arid regions) makes the solution of limestone a greater worldwide concern in pipeline route selection and characterization.

Sinkholes are mostly funnel-shaped circular depressions measuring a few metres to tens of metres across (Figure 31), formed by the downward seepage of water through residual soil overburden, enlarging fissures and cavities in the underlying soluble rock. Sinkholes develop where the watertable has lowered from above to below the soil-rock boundary, and where the rate of percolation has been accelerated by an increasing amount of water available at ground surface.

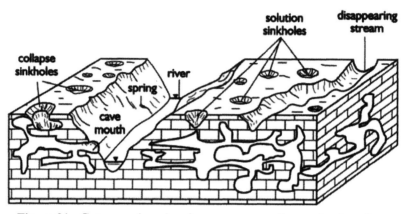

Figure 31 - Cutaway view showing common surface and subsurface dissolution features in karst terrain

The development of karst is common in tropical and semitropical carbonate and evaporite regions, where dense forests, thick organic soils and heavy rains combine to produce carbonic acid groundwater. Cone (cockpit) karst is common in the tropics, characterized by clusters of cone-shaped hills surrounded by star-shaped alluvial flats. In southern and northern (*e.g.*, arctic) arid desert terrains, where precipitation is low and the surface of carbonate terrains is dry and the organic topsoil is thin, classic karst terrain is less common.

There are two types of sinkholes in classic karst: solution sinkhole karst, caused by the dissolving action of acid groundwaters on soluble bedrock; and collapse sinkhole karst, created when the roof above underground cavities suddenly drops. As a result, most non-tropical karst topography is characterized by a pockmarked pattern of dimpled sinkholes and sparse surface runoff channels (Figure 32). The relatively few large valleys that do appear in airphotos typically have steep-walls and flat bottoms, and a common absence of tributary gullies. Streams disappear into sinkholes, connecting with underground channels and cavities following solution-widened subvertical fractures and subhorizontal bedding planes, where the carbonate strata have not been significantly folded or tilted. Groundwater entering sinkholes often reappears downstream as flowing springs, which can disappear during dry weather and reappear during wet weather.

Long linear fracture-trace lineaments are a common feature over surficial and near-surface limestone and evaporite formations, reflecting the continuity of aligned sinkholes (Figure 33) and linear solution depressions. All karst terrains display them. Ground observations reveal sinkholes and elongate solution valleys to be aligned along surface lineaments that overlie narrow zones of solution-widened master joints, joint zones, and faults of minor displacement. Sinkholes are avoided in pipeline route selection where it is practical to do so, especially in locations where they are actively enlarging and deepening. Sinkhole topography is a serious pipeline routing impediment where sinkholes are deep, steep-sided and closely spaced, thus difficult to avoid along linear pipeline routes.

Figure 32 - Pockmarked karst terrain in the southeastern United States, illustrating the difficulty of avoiding closely spaced sinkholes. D: dry sinkhole; W: water-filled sinkhole, Kentucky, USA
(United States Department of Agriculture airphoto)

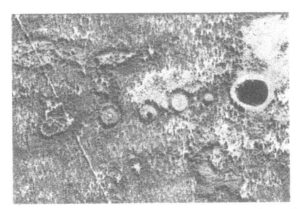

Figure 33 - Airphoto showing aligned sinkholes resulting from dissolution of evaporite strata in carbonate beds underlying thick glacial overburden, Fort McMurray oil sands area, Alberta, Canada

Karst features and their associated obstacles and hazard problems, routinely identifiable on airphotos, are listed below.

- Solution and collapse sinkholes. A typical sinkhole is roughly funnel-shaped, circular, 25 to 50 m in diameter, and 3 to 15 m in depth. Sinkholes may, however, range anywhere from 5 to 150 m in diameter

- Elongate (long and narrow) lakes that overlie solution-widened fractures in tilted carbonate strata

- Sinkholes that need to be capped or filled to minimize excessive subsidence

- Steep-sided sinkholes with and without ponds and lakes

- High relief conical "haystack- and cucumber-shaped" residual hills, separated by alluvial flats, in tropical karst terrain (cone karst), which present obstacles in pipeline route selection

- Sinkholes that flood during flash thunderstorms

- Groundwater discharge into and out of sinkholes

- Hard carbonate-quality groundwater

- Large flowing springs

- Underground cavities that are a safety hazard owing to possible roof collapse. Such collapse can leave a pipe suspended above an empty space

- Sinkholes in sand and till over carbonate beds, often caused by solution collapse of highly soluble evaporite (salt, gypsum) interbeds in carbonate strata (*e.g.*, Wood Buffalo National Park, northeastern Alberta)

- Sudden collapse cave-ins and downdrops

- A difficult-to-predict sharp, irregular contact between surficial residual soil overburden (up to 5 m deep but mostly much thinner) and underlying carbonate bedrock. Thus a pipeline trench excavation that alternates rapidly from soil to bedrock

- Highly plastic residual clay formed from dissolved limestone and subject to significant swelling and shrinking. A soft, weak residual clay often forms at the base of residual soil over carbonate rock

- Disappearing streams and reappearing springs

- Watertable surfaces that experience large elevation fluctuations, which are typically difficult to define

- Salt corrosion of pipelines

- Gypsum heave of pipelines

- Spillage of petroleum products into sinkholes and thus access to the underlying groundwater system

- Large groundwater source, although water that is susceptible to high hardness, pollution and contamination

2.4.9 Desert Terrains

A high proportion of the world's largest oil and gas fields are located in desert terrains, most notably in North Africa, southwest Africa, the Middle East, and parts of southwestern United States, southwestern Canada, Mexico, southwestern South America, southeastern Europe, southern Russia, western China and Australia (Figure 34).

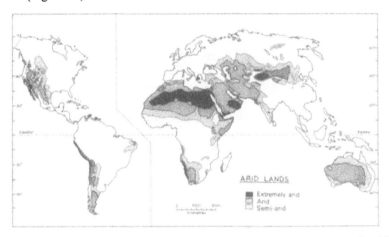

Figure 34 - Small-scale map showing extremely arid, arid and semi-arid lands, where many of the world's large oil fields are found
(*from* **Brunsden and Doornkamp, 1975**)

Deserts cover about 30 per cent of the Earth's continental surface. Much of this area consists of extensive sand sheets and sand seas, although only about 20 per cent of the world's deserts have sand dunes. In the giant dune fields in the Sahara, Middle East and South Africa, dunes can exceed 100 m high, several kilometres long with basal widths as much as a kilometre, forming long subparallel chains one to three kilometres apart. Sand seas consist of more than 40 per cent coarse sand and gravel, whereas sand dunes can have up to 80 per cent fine sand (0.1 to 0.2 mm approximately), the remainder being medium sand (0.2 to 0.6 mm). Nearly 50 per cent of all desert surfaces are wind-eroded (deflation) areas, where wind has removed the fines (silt and clay) and sand, leaving bare rock or a surface layer composed of pebbles and cobbles on top of bedrock, the landscape called a "hammada".

Deserts are common features where the average annual precipitation is arid (less than 250 mm) to semiarid (250 to 500 mm). Some desert landscapes receive as little as 110 mm a year total precipitation. Deserts are also regions of high evaporation: 800 to 4000 mm a year. As a result, they are characterized by few plants, strong winds, rare torrential rainstorms, playa lakes, salt flats, thin soil profiles, and lime-cemented subsoils called "calcrete". Dunes, dune fields, dust, playas, blowouts, wind-scour streaks and pits, alternately wind-eroded ridges and intervening trenches in soft clayey sand called "yardangs", and desert oases can be observed in airphotos and satellite imagery depending on their scale.

Many deserts are isolated, inhospitable, inaccessible regions. Strong winds create environmental effects that negatively affect pipeline construction and pipeline maintenance: dust storms, low visibility, a hot dry climate, excessive wind erosion, shifting dunes. And dust storms and heat can affect the health of pipeline workers.

There are two main types of wind erosion: deflation and abrasion. Deflation is the removal of material and abrasion is the sand-blasting or abrasive effect of wind-carried material.

Also, there are two main types of wind (aeolian) deposit: sand dunes, consisting of ridges and mounds of windblown dominantly fine and medium sand; and loess, a homogenous silt-rich, buff-coloured, loose and porous windblown deposit. Loess is estimated to mantle one-tenth of the Earth's land surface, ranging in thickness from less than 1 m to over 100 m. Notable examples are found in China, central Europe, southern Russia and northwestern and central parts of North America. High streambanks, gully sides and highway ditch backslopes in thick loessial deposits can often be observed in near-vertical slopes.

There are several common types of dunes: barchan, longitudinal, star (pyramidal), parabolic and domal (Figure 35) as well as transverse (normal to the prevailing wind direction). Each dune type has a distinctive shape, often occurring within a particular region of the world and physical environment and influenced by the availability of sand supply, wind strength, annual moisture and thus vegetation cover. Parabolic dunes (U-shaped dunes with horns pointing upwind) are a common dune formed during the Pleistocene under more moist periglacial climatic conditions. One of the more impressive dunes is the long and narrow longitudinal (seif) dune, where dune crests parallel the dominant wind direction, thought to be formed by counter-rotating turbulent roles of spiral wind vortices. Some dunes are "fixed" by vegetation while others are sheltered by topographic relief, and still others are constantly reshaped and shifted around by the wind.

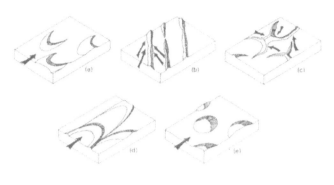

prevailing wind direction(s)

Figure 35 - Five dune types: a) barchan, b) longitudinal, c) star (pyramidal), d) parabolic, e) domal

Following is a list of landforms, terrain obstacles, geohazards and other problems characteristic of desert regions, where today many long, large diameter pipelines are located and major decisions are required on pipeline route selection and characterization:

- All dune types, including modifications of their more common characteristic shapes, consisting largely of fine-grained sand

- Active dunes including sand in blowouts that expose pipeline coatings to wear

- Desert dunes that march downwind and expose pipelines

- Clean, loose sand on downwind steep (~34°) lee-slopes subject to slipping, where wheeled vehicles are easily bogged down. This is because the void ratio, porosity and looseness of sand are greater on lee slopes than on windward slopes that are impacted, compacted and densified by strong winds.

- Drifting fine sand that gets into and accelerates wear on pipeline construction and maintenance machinery

- Areas where clean fine sand is dry and where it is saturated in near-surface watertables, both increasing the likelihood of trench walls caving and therefore endangering workers

- Dust storms that pose a health and safety hazard, largely caused by severely reduced visibility

- High turbulence effects created by high winds blowing around topographic obstacles

- Lee slope sand slumps that expose the pipe
- Windswept bare rock and surfaces veneered with gravel and cobble lag deposits
- Deep deposits of loess, especially where denuded of vegetation, that are susceptible to excessive gully erosion
- Collapsible soil structure from wetting of *in situ* loess
- Collapse from piping in *in situ* loess and redeposited colluvial loess (Figure 36)
- Ground subsidence where overpumping of desert aquifers lowers the watertable
- Ground settlement from desiccation, exposing the pipe
- Loose loess susceptible to sudden subsidence from earthquake ground-shaking
- Deep near-vertical cuts in thick loess deposits (Figure 37)
- Alkali lakes and saline soils that induce pipeline corrosion
- Calcrete deposits, consisting of lime-cemented crusts, that can present a difficult-to-excavate hardpan
- Rare torrential rains resulting in flash floods that divert distributaries on alluvial fans, leaving bridges abandoned and culverts filled with sand and gravel debris. New channels form and existing ones are abandoned
- Flash floods that cause washouts, exposing the pipe
- Ephemeral sand and gravel streambeds—variously called arroyos, wadis and nullahs—which are typical desert watercourse features and channels subject to migration
- Alluvial fans in desert terrains
- Hard saline crusts occurring over a very soft unstable soil deposit
- Saline soils and playas that cause pipeline corrosion
- Desert venomous snakes and prickly plants, human hazards

These surface features and conditions are things the remote sensing analyst should keep in mind when locating and characterizing terrain and geohazards in pipeline route selection.

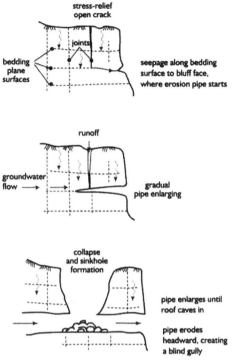

Figure 36 - Development of piping and roof collapse in thick silty waterlaid (fluvial, lacustrine, marine) and silty windlaid (loessial) deposits, characteristic of extremely arid, arid and semi-arid lands

Figure 37 - Near-vertical cuts, typical of thick loess deposits in Nebraska, USA

2.5 Synthesizing Terrain Datasets within a GIS

The advent of Geographic Information Systems (GIS) has greatly improved the ability of pipeline engineers and geoscientists to collect, standardize, synthesize and analyze geological, geotechnical and socioenvironmental datasets considered when they are locating and planning pipeline routes. This is particularly true for long pipeline routes that span several jurisdictions, where differences in scale, collection methodology and data presentation between jurisdictional maps can make traditional "desktop" data synthesis difficult if not impossible. As long as the datasets contain a proper geographic reference (*i.e.*, each data point in the dataset is assigned to a specific geographic point on the Earth's surface), virtually all information of importance to pipeline route selection and charactization can be incorporated into a GIS.

Existing digital datasets can be imported directly into a GIS (Figure 38). However, datasets that exist only in paper map or text format must be converted into digital format before incorporation into a GIS. One of the simplest way to do this is scanning the maps with a large format digital scanner, and importing the resulting raster images directly into a GIS. Once geocorrected, these scanned maps can be displayed along with other data in the GIS; however, terrain analysis is limited to direct visual comparisons only. A more advanced and useful method is to capture map information as vector points, lines and polygons with attached attribute data. This is done through manual digitization of map features or using automated raster-to-vector conversion processes, or a combination of both. Capturing the map data in vector format allows for a more advanced synthesis and analysis in the GIS environment, such as automatic querying and display of specific terrain features of interest in route planning, and the generation of ancillary data drawn from multiple datasets. But because conversion of a series of paper maps or text datasets into digital form can be expensive and time-consuming, planners must balance the cost of digital dataset capture against the potential value of the dataset in the route planning process. In some cases, partial digitization of the dataset may be sufficient for planning requirements.

Once all applicable datasets have been incorporated into the GIS, data can quickly be displayed and analyzed to identify alternative pipeline route locations, and also to identify terrain and geohazard constraints to route selection. The ability to generate ancillary datasets from existing data is one of the most powerful features of a GIS, and one of the most useful to pipeline route planning and construction engineers. For example, data from a soil survey map can be combined with slope and aspect data from a digital elevation topographic model to identify areas at risk of slope failure—and even assess the degree of risk depending on the nature of several interacting variables. Proposed pipeline routes can be overlain on a risk map, highlighting potential problem areas along the route, suggesting competing routes or identifying segments of problem route to avoid.

Figure 38 - Importing existing paper geospatial data into a Geographic Information System (GIS)

GIS software can also be used to display spatial relationships between constraining factors that influence route planning. For example, during stakeholder meetings it may be determined that a pipeline route is not permitted to pass within a certain distance of a protected area, such as a park or wilderness reserve, with the exact distance determined by the administrative and legislative classification of the protected area. Using the spatial editing tools in a GIS, pipeline engineers can automatically generate buffers of the appropriate width around the boundaries of the protected areas and display these buffer zones as a thematic layer in the GIS. The end goal of these GIS-supported terrain and hazard analyses is to present as much information as needed by the route engineer to make effective decisions in route selection—and in the quickest and most efficient way possible.

Today in terrain and geohazard mapping and analysis studies, provisional route corridors, typically about two kilometres wide, are often identified initially from information appearing on topographic, soil, geology, landuse, landcover and environmental maps and satellite imagery. Corridor alternatives are then rated on the basis of cost, engineering and other factors (social, cultural, political, regulatory, etc.), and the most competitive routes selected for in-depth airphoto terrain analysis. Such detailed analysis is carried out using stereoscopic airphoto prints or stereoviewing computer software. In either case, the result is a multidiscipline-selected preferred pipeline route right-of-way (ROW). The ROW centreline location is then commonly entered into a GIS for integration with other datasets, further analysis and presentation to project proponents and stakeholders. Airphotos used for detailed route selection are scanned and imported as georeferenced raster images into the GIS.

Once provisional route corridors and detailed route alignments are identified, the GIS is used to display the route locations against a background of geoengineering and environmental data for discussion in public consultations and stakeholder meetings. The route is typically displayed on a satellite mosaic of the route study area, with features of interest and constraining factors displayed as thematic layers. These can be turned on or off depending on the issues being discussed. In some cases, modifications or additions to a provisional route can be made during the time of discussion, ensuring a permanent record of changes that satisfy participants, so far as possible, in the route selection and characterization process.

These are GIS datasets commonly used in pipeline route selection studies:

Raster data

- Satellite images
- Airphotos
- Scanned topographic and thematic maps

Vector data

- Surficial geology maps
- Soils maps
- Environmental data maps
- Infrastructure maps

Point data

- Facilities
- Buildings and residences
- Wells
- Field observations
- Ground photographs
- Digital elevation data points

2.6 Presenting Pipeline Terrain Data in a GIS

With a final pipeline route established, there is need to present a preferred route and competing alternatives to the project proponent and stakeholders to review the route selected and its characterizing features and controls. In these meetings, presentation of the route location and its terrain characteristics may take the form of printed maps of varying sizes and themes. Alternatively, the route selection and

60

characterization information can be presented in digital form within a GIS environment. Presenting the route data in a GIS format has the advantage that large amounts of data of interest to stakeholders can be included in the GIS database as thematic layers, which can be turned on and off as necessary depending on the issues being discussed.

Typical route location and characterization datasets that may be included in a GIS-based route presentation are engineered pipeline data (location of compressor stations and valves, or physical line characteristics such as size, coating, weld type, etc.); airphoto-mapped terrain and geohazards data; field geotechnical, geological and hydrogeological data, topography, drainage, vegetation and environmental data (such as protected places); and socio-economic and politico-cultural data—all of relevance to the route selection and characterization process.

Each of these data types can be displayed on a computer screen in any combination in the selection process, to show relationships between the final route selected and various constraining or controlling factors. Stakeholders in these meetings can manipulate different layers of information to display features of significance to the discussion and to highlight important points about the route selection process. For stakeholder meetings with large groups of participants, digital projectors and large-screen video displays can be used to show the GIS presentation in a format easily viewed by everyone. As well, print versions of specialized maps can be readily generated for any stakeholder group.

To further enhance planner and stakeholder understanding of the preferred route selection and characterization process, interactive 3-D technology similar to that used by pipeline engineers and others may be used to produce realistic three-dimensional fly-overs of the selected route. Satellite imagery used in the planning process can be draped over three-dimensional surfaces generated from digital elevation models to produce a simulation of the topography occurring along the route. Identified routes and characterization data can be draped as additional layers over this 3-D landscape. Data users can "fly" through this virtual landscape as if they were in a helicopter or fixed-wing aircraft, shifting around the landscape wherever and however they choose, even stopping in mid-flight to examine a particular terrain feature or adverse ground condition. Most GIS packages currently on the market have a 3-D fly-through process, either inherent in the program or as an add-on feature. With high-resolution full-colour imagery, rendered sky textures, independent lighting and haze simulators and the powerful graphic capabilities of modern computers, these 3-D fly-throughs can be stunningly realistic (Figure 39).

Figure 39 - GIS-rendered 3D flythrough along pipeline route

Acknowledgements

The authors wish to thank Moness Rizkalla, pipeline engineering and integrity management specialist, Visitless Integrity Assessment Ltd, for the invitation to prepare this chapter; Charles R. Neill, Northwest Hydraulics Ltd, for suggestions on Figure 14 and for the top and center series of sketches on river scour in Figure 18; and David Kerr, Golder Associates, for papers on the environmental impact assessment of large pipeline projects. The authors also wish to thank David Boschman, J.D. Mollard and Associates Ltd, for formatting the text and 39 figures. Airphotos in Figures 7, 8, 10, 13, 14, 24, 26 and 33 are reproduced courtesy of National Resources Canada, National Air Photo Library.

References and Additional Reading

(1) Arthur, J.C.R., Haas, C., Shilston, D.T., and Waltham, A.C., "The Sivas Karst—from terrain evaluation to ground truth." International Conference on Terrain and Geohazard Challenges Facing Onshore Oil and Gas Pipelines: Institution of Civil Engineers, London, UK, 2005

(2) Belcher, D.J., "Identifying landforms and soils by aerial photographs," *Proceedings, 39th Annual Purdue Road School,* 1944

(3) Belcher, D.J., *Aerial Photographic Reconnaissance Investigation—A Permafrost Study in the Territory of Alaska*, U.S. Army Corps of Engineers, St. Paul District Office, December, 1945

(4) Belcher, D.J., "Determination of soil conditions by aerial photographic analysis," *Proceedings, Second International Conference on Soil Mechanics and Foundation Engineering*, vol. 1, pp. 313-321, 1948

(5) Bird, J., O'Rourke, T., Bracegirdle, T., Bommer, J., and Tromans, I., "A framework for assessing earthquake hazards for major pipelines," International Conference on Terrain and Geohazard Challenges Facing Onshore Oil and Gas Pipelines: Institution of Civil Engineers, London, UK, 2005

(6) Brunsden, D., "Geohazards and pipeline engineering," keynote paper, International Conference on Terrain and Geohazard Challenges Facing Onshore Oil and Gas Pipelines: Institution of Civil Engineers, London, UK, 2005

(7) Brunsden, D. and Doornkamp, J. C., "The Unquiet Landscape," a Halsted Press Book published by John Wiley & Sons, Inc. New York. Chapter 21 on Deserts, p 126, 1975

(8) Charman, J.H., Fookes, P.G., Hengesh, J.V., Lee, E.M., Pollos-Pirallo, S., Shilston, D.T., and Sweeney, M., "Terrain, ground conditions and geohazards: evaluation and implications for pipelines," International Conference on Terrain and Geohazard Challenges Facing Onshore Oil and Gas Pipelines: Institution of Civil Engineers, London, UK, p. 81, 2005

(9) Fookes, P.G., Lee, E.M., and Sweeney, M., "In Salah Gas project, Algeria— part 1: terrain evaluation for desert pipeline routing," International Conference on Terrain and Geohazard Challenges Facing Onshore Oil and Gas Pipelines: Institution of Civil Engineers, London, UK, 2005

(10) Frost, R.E., *Evaluation of Soils and Permafrost Conditions in the Territory of Alaska by Means of Aerial Photographs,* for the Office of the Chief of Engineers, Airfields Branch St. Paul District, U.S. Army Corps of Engineers; volumes 1 and 2, 1950

(11) Guthrie, R.H., and Shilston, D.T., "Hazards in moutainous terrain: lessons for linear infrastructure from the Canadian forest industry," International Conference on Terrain and Geohazard Challenges Facing Onshore Oil and Gas Pipelines: Institution of Civil Engineers, London, UK, 2005

(12) Hearn, G.J., "Thirty years of terrain evaluation for route corridor engineering," International Conference on Terrain and Geohazard Challenges Facing Onshore Oil and Gas Pipelines: Institution of Civil Engineers, London, UK, p. 206, 2005

(13) Hoyt, J.B., *Man and the Earth*, Prentice Hall, Inc. Englewood Cliff, New Jersey, USA, p. 99, 3rd edition, 1973

(14) Kerr, D., "Environmental impact assessment of large pipeline projects in remote environments: integrating engineering and environmental design," International Conference on Terrain and Geohazard Challenges Facing Onshore Oil and Gas Pipelines: Institution of Civil Engineers, London, UK, 2005

(15) Kreig, R.A., and Metz, M.C., "Recent advances in route selection and cost estimating methodology in Arctic and Subarctic regions," in International Symposium on Geocryological Studies in Arctic Regions, Yamburg, U.S.S.R., 1989

(16) Krieg, R.A., and Reger, R.D., "Preconstruction terrain evaluation for the Trans Alaska pipeline project," in Coaste, D.R. (Ed.), Geomorphology and Engineering, 7th Annual Geomorphology Symposium, New York, pp. 55-76, 1976

(17) Lee, E.M., and Charman, J.H., "Geohazards and risk assessment for pipeline route selection. Terrain, ground conditions and geohazards: evaluation and implications for pipelines," International Conference on Terrain and Geohazard Challenges Facing Onshore Oil and Gas Pipelines: Institution of

(18) Lotter, M., Charman, J.H., Lee, E.M., Hengesh, J.V., Shilston, D.T., and Poscher, G. "Geohazard assessment of a major crude oil pipeline system—the Turkish section of the BTC Pipeline route," International Conference on Terrain and Geohazard Challenges Facing Onshore Oil and Gas Pipelines: Institution of Civil Engineers, London, UK, 2005

(19) Lukas, A., Loneragan, S., and MacDonald, D., "The practicality of drilling very long pipelines under hazardous terrain—5 km, 10 km?" International Conference on Terrain and Geohazard Challenges Facing Onshore Oil and Gas Pipelines: Institution of Civil Engineers, London, UK, 2005

(20) Manning, J., Willis, M., Denniss, A., and Insley, M. "Remote sensing for terrain evaluation and pipeline engineering," International Conference on Terrain and Geohazard Challenges Facing Onshore Oil and Gas Pipelines: Institution of Civil Engineers, London, UK, 2005

Project Success: Addressing "Feeling/Perception" Issues" " Proc. ASME-OMAE 2nd International Pipeline Conference (IPC), Palliser Hotel, Calgary AB, Canada, Vol 1, pp 95-101, 1998

(22) Mollard, J.D., "Photo Interpretation of Transported Soil Materials," *The Engineering Journal,* Engineering Institute of Canada, vol. 32, no. 56, pp. 332-340, 1949

(23) Mollard, J.D., "How to Use Aerial Photos in Pipelining," *Pipe Line Industry Magazine*, Gulf Publishing Company, Tulsa, Okla., 1959

(24) Mollard, J.D., "Photo Interpretation in Engineering" (Chapter 6), *Manual of Photographic Interpretation*, American Society of Photogrammetry, 1960

(25) Mollard, J.D., "Guides for the Interpretation of Muskeg and Permafrost Conditions from Aerial Photographs," *Oilweek*, Calgary, Alberta; July issue, 1960.

(26) Mollard, J.D., "Photo Analysis and Interpretation in Engineering Geology Investigations: A Review," *Reviews of Engineering Geology*, Vol. 1., Geological Society of America, ed. T. Fluhr and R.F. Legget, 1962

(27) Mollard, J.D., "The Role of Photographic Interpretation in Northern Route and Site Surveys," Sixty-first Annual Meeting, Canadian Institute of Surveying, 1968

(28) Mollard, J.D., "Airphoto Location of Alternative Pipeline Routes and Associated Terrain Mapping, Mackenzie Valley Pipeline Prudhoe Bay, Alaska, to Northwestern Alberta"; Volume 1, Route Location and Terrain Mapping; Volume 2, Terrain Legend, *Northwest Project Study Group and Mackenzie Valley Pipeline Research*, 1971

(29) Mollard, J.D., "Airphoto Terrain Classification and Mapping for Northern Pipeline Feasibility Studies," *Proceedings Canadian Northern Pipeline Research Conference*, Tech. Mem. 104 (NRCC 12498), Ottawa, 1972

(30) Mollard, J.D., "Airphoto Interpretation: Fluvial Features," Ninth NRC Canadian Hydrology Symposium on Fluvial Processes and Sedimentation, Edmonton, Proceedings of Conference, National Research Council, 1973

(31) Mollard, J.D., "Regional Landslide Types in Canada," *Reviews in Engineering Geology*, Vol. III, Part 1, Overview, edited by Donald R. Coates, Geological Society of America, Boulder, Colorado, 1977

(32) Mollard, J.D., "Contribution to Site and Route Studies" in Chapter 5, "Permafrost," *Engineering Design and Construction*, Associate Committee on Geotechnical Research, edited by G.H. Johnson, National Research Council of Canada, Ottawa, 1981

(33) Mollard, J.D., "Development of Airborne and Satellite Remote Sensing Applications," *Proceedings of the Seventh Canadian Symposium on Remote Sensing*, Winnipeg, eds. William G. Best and Sue-Ann Weselake, 1981

(34) Mollard, J.D., *Landforms and Surface Materials of Canada: A Stereoscopic Airphoto Atlas and Glossary*, 8th edition, J.D. Mollard and Associates Ltd., Regina, Saskatchewan, 1982

(35) Mollard, J.D., "Environmental Factors and Interrelationships that Guide Engineering Photointerpretation of Canadian Peatlands," Symposium on Advances in Peatlands Engineering, sponsored by National Research Council of Canada, Carleton University, Ottawa, Ontario, 1986

(36) Mollard, J.D., *Down to Earth: Applied Multidiscipline Remote Sensing for Natural Resource and Infrastructure Investigations*. J.D. Mollard and Associates Ltd., Regina, Saskatchewan, 1995

(37) Mollard, J.D., and Pihlainen, J.A., "An Application of Photo Interpretation to Road Selection in the Arctic," First Internation Conference on Permafrost, NAS-NRS Publ. 1287, pp. 381-387, 1963

(38) Mollard, J.D., and Janes, J. Robert, *Airphoto Inerpretation and the Canadian Landscape*, Energy, Mines and Resources Canada, Canadian Government Publishing Centre, Supply and Services, Ottawa, Canada, 1984

(39) Morgan, V., Clark, J., and Hawlader, B., "Modelling of frost heave of gas pipelines in Arctic conditions," International Conference on Terrain and Geohazard Challenges Facing Onshore Oil and Gas Pipelines: Institution of Civil Engineers, London, UK, 2005

(40) Morgan, V., Kenny, S., Power, D., and Gailing, R., "Monitoring and analysis of the effects of ground movement on pipeline integrity," International Conference on Terrain and Geohazard Challenges Facing Onshore Oil and Gas Pipelines: Institution of Civil Engineers, London, UK, 2005

(41) Nadim, F., and Lacasse, S., "Mapping of landslide hazard and risk along the pipeline route," Terrain, ground conditions and geohazards: evaluation and implications for pipelines. International Conference on Terrain and Geohazard Challenges Facing Onshore Oil and Gas Pipelines: Institution of Civil Engineers, London, UK, 2005

(42) Neill, C.R., and Mollard, J.D., "Examples of Erosion and Sedimentation Processes Along Some Northern Canadian Rivers," UNESCO-sponsored International Symposium on River Sedimentation, Beijing, China, 1980

(43) O'Connell, P., Boyd, G., Insley, M., and Murphy, P., "Pipeline engineering team's response to challenges of terrain and geology," International Conference on Terrain and Geohazard Challenges Facing Onshore Oil and Gas Pipelines: Institution of Civil Engineers, London, UK, 2005

(44) Penner, L.A., Mollard, J.D., and Stokke, P., "Modelling the shore erosion process for predicting bank recession around Canadian Prairie and northern permafrost-affected lakes and reservoirs," in *Erosion: Causes to Cures*, Short Course and Conference, Canadian Water Resources Association, Soil and Water Conservation Society and International Erosion Control Association, Regina, Saskatchewan, 1992

(45) Péwé, T.L., "The Periglacial Environment" Arctic Institute of North America: McGill-Queen's University Press, Montreal, 1969

(46) Porter, M., Esford, F., and Savigny, K.W., "Andean pipelines—a challenge for natural hazard and risk managers," International Conference on Terrain and Geohazard Challenges Facing Onshore Oil and Gas Pipelines: Institution of Civil Engineers, London, UK, p. 262, 2005

(47) Savigny, K.W., and Isherwood, A.E., "Terrain and geohazards challenges to pipelines in Canada," International Conference on Terrain and Geohazard Challenges Facing Onshore Oil and Gas Pipelines: Institution of Civil Engineers, London, UK, 2005

(48) Shilston, D.T., Lee, E.M., Pollos-Pirallo, S., Morgan, D., Clarke, J., Fookes, P.G., and Brunsden, D., "Terrain evaluation and site investigations for design of the trans-Caucasus oil and gas pipelines in Georgia," International Conference on Terrain and Geohazard Challenges Facing Onshore Oil and Gas Pipelines: Institution of Civil Engineers, London, UK, 2005

(49) Snead, R.E., "Atlas of world physical features," John Wiley & Sons, Inc., map on volcanic features p. 55, 1972

(50) Sweeney, M., "Terrain and geohazard challenges facing onshore oil and gas pipelines: historic risks and modern responses," International Conference on Terrain and Geohazard Challenges Facing Onshore Oil and Gas Pipelines: Institution of Civil Engineers, London, UK, 2005

3 Open Cut and Elevated Pipeline River Crossings

3.1 *Introduction*

3.1.1 General

Historically, the frequency of pipeline exposures and failures at river crossings has exceeded that of overland pipelines. Major constructed and proposed projects, especially those in extreme environments in the Arctic or the rugged terrain of the Andes in South America, have evolved the science of river crossing design. Extensive regulatory reviews of major projects have also influenced the design process significantly. Government mandated operational monitoring requirements and the owners Best Management Practices re: maintenance has, in combination with the more rigid design process and improved quality control during construction, reduced integrity concerns associated with river crossings.

An increasing challenge in the last several decades is the environmental restrictions associated with instream construction. This has lead to significant improvements in the development of trenchless crossing techniques such as horizontal directional drilling (HDD). Nevertheless for economic or technical reasons, open cut crossings, with or without flow isolation during construction, will remain a dominant river crossing construction method. To a smaller degree, except for unique cases such as a hot-oil pipeline in the Arctic, elevated river crossings, ranging from simple pipe spans to long suspension bridges, are alternative methods of construction.

The multitude of pipelines in a single corridor in some parts of the world tend to result in less and less favourable river crossings over time – the best river crossing alignment generally is selected for the first pipeline. This poses additional technical and environmental challenges for subsequent pipelines.

3.1.2 Approach to and Use of this Chapter

The objectives of this chapter are to outline the factors and steps to be considered and undertaken in the design, construction and operation of pipeline river crossings. Hydrologic, hydraulic and river engineering procedures and techniques are outlined in a general and simplified manner – numerous standard texts and design manuals are available and present detailed methodologies.

The purposes may therefore be summarized as follows:

- to serve as a primer for a junior design engineer responsible for the design of river crossings;

- to provide an overview of the key river crossing issues and considerations for the pipeline project manager or regulator; and
- to, via numerous photos, illustrate examples of typical and unique design solutions and construction techniques. This will be useful for all project personnel and stakeholders.

The importance of or lack of importance of certain design steps and analysis are stressed throughout this document. For example, the quantative hydrologic analysis to compute the design flow for buried crossings is generally not critical whereas the qualitative assessment of the potential for local scour to form at a crossing is often the key to the sound and economic design of the crossing.

3.1.3 Organization

The sequence presented in this chapter follows the usual project sequence in the planning, design, construction and operation of pipelines.

Section 3.2 details the field data needs, design steps and design methodologies. Unique design challenges, such as debris flows and glacier dammed lakes are discussed briefly and illustrated as well as other unique solutions such as instream pipeline alignments or parallel alignments proximate to the rivers are discussed. Typical designs for buried and elevated crossings are presented.

Section 3.3 presents open cut construction techniques with and without flow isolation. Numerous examples are illustrated for both via photographs and schematics for both arctic and temperate conditions.

Section 3.4 is a brief discussion of the importance of monitoring major river crossings and if necessary, constructing additional protective works. Examples of summary monitoring sheets, to annually update the condition report of the crossing, are presented.

Section 3.5 lists the references quoted in this document. Standard technical textbooks and guides in hydrology, hydraulics and river engineering, non-specific to pipeline river crossings, are not listed.

3.2 *Design*

3.2.1 General

This section presents the required design steps, criteria and methodologies for buried and elevated pipeline river crossings. The objective of the design process is to produce a sound, practical and economic design that, for the life of the pipeline, minimizes the risk of pipeline exposure and failure for the design flood and for foreseeable future river conditions.

The design of crossings involves quantitative analysis and qualitative judgements. Whereas some components of the design can be quantified with adequate data (design flood and water level for example) others cannot be (potential future river changes for example). As discussed and illustrated in this section, it is generally the qualitative considerations that are critical to a sound, practical and economic design of major river crossings. Qualitative judgements should be based on extensive river crossing design and operational experience.

3.2.2 Quantitative Versus Qualitative Analysis

Veldman (1983) stressed the importance of the qualitative component of pipeline design:

"The environmental and security-of-supply concerns for recent major projects have resulted in design stipulations requiring non-exposure of the pipeline during extreme flood events for the life of the project. The United States Federal and State review bodies, established for projects such as Trans-Alaska Pipeline and Northwest Alaska Gas Pipeline, have necessitated detailed hydrologic and river engineering studies, analysis and documentation – a first in the pipeline history. This has resulted in the development of novel and extensive techniques, approaches and mathematical models to quantify the hydrological, hydraulic and river engineering conditions.

The design pendulum for recent major projects appears to be swinging too far in the direction of trying to develop elegant mathematical models. Designers are expounding various models to predict scour and even bank migration and in their attempts to find elegant mathematical solutions – and thus "improve" on methodologies such as the Blench regime method – overlook the fact that many of the river engineering processes cannot be quantified. Furthermore, the mathematical models often require the same type of "fudge factors" and assumptions as the often-criticized empirical methods. This desire for quantitative and precise solutions may, to a degree, be attributed to governmental review requirements. There is no question that quantitative analyses are easier and quicker to review, whereas qualitative ones, depending on the experience of the designer and reviewer, may, at times, lead to long debates, the end result of which is a compromise which satisfies neither party. It is not a pure science but mainly an art.

This desire or need to produce quantitative techniques and results can also result in a disproportionate amount of the designer's and reviewer's time spent on some of the less important aspects from an integrity viewpoint. For example, computing the precise magnitude of the design flood for a buried crossing of a braided river or its burial depth – two factors which can be quantified to a degree – is less important from an integrity viewpoint than assessing the potential for channel switching or a subchannel development – factors which cannot be quantified. Also an inherent danger in mathematical models is to analyze only the immediate crossing area. With qualitative approaches, the behavior of the

71

river over a much longer reach can be assessed. Nanson and Hickin (1983) have identified the shortcomings of predicting migration rates for specific areas from data at individual bends."

3.2.3 Integration with Other Disciplines

The hydrotechnical design of the crossings, primarily the focus of this section, must be integrated with geotechnical and pipeline design, and environmental and construction considerations and limitations. For example:

- The determination of the preliminary pipe profile should be done in coordination with geotechnical engineers and pipeline design and construction specialists. The optimum sagbend setback may be governed by slope stability, thermal or pipe bending or construction considerations rather than river engineering reasons. Bank protection structures therefore may be necessary, not for river engineering reasons, but for geotechnical and constructability reasons to reduce excavations into steep, high banks.
- The design should be reviewed and coordinated with environmental specialists. Environmental considerations may dictate or influence the crossing design. For example bank protection structures may not be desirable or may not be permitted in critical aquatic habitat areas or could in fact be favoured to enhance aquatic resources.

The appropriate design steps for major river crossings are:

- A multi-discipline start-up meeting with design, construction and operations personnel to discuss the design basis, criteria and philosophy of the project, alignment considerations and limitations, operational considerations (access, practicality of constructing repair measures, if necessary) and construction considerations (summer versus winter, environmental timing and methodology restrictions).
- Preparation of the conceptual design of the crossing specifying the pipeline elevation, sagbend locations (in the case of buried crossings) or the scour elevation for piers/piles and the minimum length and height of bridges (in the case of elevated crossings).
- Detailed design of the buried crossing specifying pipe bending and weighting requirements. For steep and high banks, the sagbend locations may need to be modified from the conceptual design to avoid excessive excavation into the bank. This may require bank structures, such as rock armor, to protect the sagbend. For elevated crossings, the bridge design would be detailed.
- Review and approval of the detailed design by the hydrotechnical design engineer. If bank structures are needed to accommodate the detailed

pipeline profile or to protect the elevated crossing, the hydrotechnical design engineer will prepare the design and construction specifications for the river related works.

3.2.4 River Classification System

Various classification systems may be utilized for environmental (fish or non-fish bearing and overwintering or non-overwintering streams for example) or construction (separate river crossing crew or part of the overland pipeline spread for example) reasons. From a hydrotechnical design viewpoint, the complexity of the crossing and its required design components produce the following system, which may or may not coincide with other classifications systems.

- **Major River Crossings** - warrant a detailed design report and design drawing. A specific size of river is not usually associated with this class but rather the complexity and design and constructions needs of the crossing. Therefore a relatively small crossing with a wide floodplain subject to change or complex alluvial fan could warrant a Major classification particularly if river training structures are required.

- **Intermediate River Crossings** - do not warrant a detailed design report or a separate detailed design drawing. Design specifications such as cover depth (or pipe elevation) and sagbend setback are generally specified for each crossing on a "generic" table of crossing dimensions.

- **Minor or Typical River Crossings** - a typical drawing illustrating minimum cover depth and pipe profile characteristics (single or double sagbends) is adequate.

3.2.5 Design Flood Criteria

A comparison of Design Flood Criteria for major projects and jurisdictions is presented in Table 1. For gas pipelines, 50 to 200 year floods are commonly used. The availability of hydrologic data is an influencing factor – for the Trans Alaska pipeline in 1973, the design rainfall value was doubled in the data-scarce Arctic slope area north of the Brooks Range.

With respect to the burial depth of pipelines at river crossings, the design scour depth is often not particularly sensitive to the flood magnitude and thus the selected design flood criteria. For example, for a crossing of a wide braided river with low banks, the hydraulic characteristics (flow depth and velocity and thus scour) increase little when the flow is overbank. Scour at bends or channel confluences, defined as local scour, is almost always more severe than general scour. At flows at or below bankfull, the flow curvature and strength of secondary currents, which cause the local scour, are generally more pronounced than at high flows which fill and parallel the entire valley.

Table 1 - Typical Flood Return Periods Used for Pipeline River Crossings

LOCATION/PROJECT	CRITERIA
1. Canadian Arctic Gas Pipeline Systems, Canada (Proposed in 1974)	In his Enquiry Report, Justice Berger proposed the Standard Project Flood being equal to about one-half the Probable Maximum Flood, the largest flood that is considered physically possible if all flood producing factors (snowmelt, antecedent conditions and rainfall) were to combine.
2. Norman Wells Oil Pipeline, Canada (completed, early 1980s.)	1:1000 year flood
3. Alberta Code of Practice for Pipelines, Canada (2001)	Pipelines carrying a non polluting medium (sweet gas) are designed for the 1:50 year flood. For pipelines conveying a polluting medium (oil, liquids and sour gas), the design is the 1:100 year flood.
4. British Columbia (current)	1:100 to 1:200 year flood
5. TransAlaska Oil Pipeline (Completed in 1977)	The Standard Project Flood as defined by the U.S. Army Corps of Engineers. The design precipitation was selected as 50% or 100% of the Probable Maximum Precipitation depending on the availability of flow data with the higher value specified for the data-scare area north of the Brooks Range.
6. Northwest Alaska Gas Pipeline (proposed in mid-1980)	The Standard Project Flood design value was compared with the 200 year estimate with the higher value governing the design flow.

74

7.	Badami Oil Pipeline, Prudhoe bay area, Alaska (completed in 1998)	The 200 year flood, as determined from a regional hydrologic analysis was utilized by regulatory agencies in their review of the design.
8.	Alpine Project, Colville River Crossings, Prudhoe Bay area, Alaska (completed in 1999).	The 100 year flood is the minimum design criteria for the design of the oil pipeline and road crossings and to establish the elevations of facilities in the delta of the Colville River.
9.	Ecuador, Argentina, Chile and Peru, South America (1997-2003)	Design flood criteria ranged from 100 to 1000 years for gas and oil pipelines.

The design flood criteria selection probably has the greatest impact on the design and cost of the floodplain fringe areas of the river rather than the main channel crossing itself. Whereas for a 50 year design, extreme and unlikely channel changes would likely not be considered during a typical 20-50 year economic life of a pipeline, utilizing a much greater criteria such as the Standard Project Flood criteria could dramatically increase the estimated zone of influence of the river and consequently require either a much longer deep buried section or more extensive river training structures to protect shallow buried line. In summary, the impact of the flood magnitude on the crossing design is site specific.

In the design of bank protection, river training structures or elevated crossings, the magnitude of the design flood and the resultant water level and velocities may have a significant impact on the design of the crossing.

3.2.6 Overview of the Design Steps and Considerations

The aim of the design[1] for buried river crossings is to minimize the risk of pipeline exposure or damage to the pipeline for the design flood and for the economic life of the pipeline. This is done via the following steps:

- computing the design flow and water level;
- assessing the potential for and magnitude of the general and local bed scour;
- analyzing the potential for bank erosion or channel switching or the development of new channels across or near the pipeline; and
- if necessary, designing bank and bed structures to protect the pipeline from the influence of the river.

The design of elevated river crossings requires an assessment of:

- the design flood magnitude, the design water level and the necessary freeboard or clearance;
- scour potential for instream piers or abutments;
- ice levels and ice breakup forces for the piers; and
- potential for channel changes and the need for and design of river training structures for the piers or abutments.

The design of river crossings involves both quantitative and qualitative assessments. The quantitative analysis consists of computing the magnitude of the

---

[1] Unless otherwise indicated, design in this report encompasses all hydrotechnical aspects such as hydrology, hydraulics and river engineering pertaining to buried crossings (pipeline,depth, sagbend and overbend setback, bank protection, river training structures) or elevated crossings (length, height of crossings, pier and abutment scour, river training structures).

76

design flood and its corresponding water level. Scour or riverbed lowering is computed via quantative techniques and qualitative assessments. The assessment of potential future bank erosion or channel switches is primarily a qualitative assessment. Depending on the magnitude of floods experienced during the available period of airphotos, historic bank erosion, as per comparative airphotos, is at times not a sound indicator of potential future erosion. Channel changes due to major floods, is generally dramatic and may not be discernable from historic changes if a major flood has not occurred during the period of comparative airphotos.

It should be noted that for smaller rivers and especially low velocity conditions, exposure of a buried pipeline does not necessarily result in pipeline integrity concerns. The aim of the design should however always minimize, for the design flood, foreseeable river conditions, and for the life of the pipeline, the probability of and extent of pipeline exposure.

The typical design steps for buried and elevated crossings are illustrated on Figure 1 and Figure 2, respectively, which illustrate the key components in a schematic fashion. The steps are:

- Measure the drainage area at the crossing from topographic plans or obtain from published information by government agencies.
- Establish the appropriate design flood criteria as established by the owner and/or by the government regulators.
- Compute the design flood from recorded streamflow data and extrapolated or interpolated to the river crossing location. For streams with no flow data, compute the design flow from regional data in the same hydrologic zone or compute the flow from rainfall data and an appropriate hydrologic model. In order to calibrate the rainfall-runoff model, a historic major rainfall event and the corresponding riverflow, must be available.
- Compute the design water level by extrapolating historic high water marks from known water levels at bridges, for example, and/or by computation using river surveys (at least three cross sections required). Computations can be done via a simple backwater analysis or by commonly used models such as HEC-RAS developed by the U.S. Army Corps of Engineers. The impact of ice (breakup and icings) on the design water level must be recognized. For many northern rivers, a moderate flow during breakup produces the maximum water level.
- Measure or estimate the surficial riverbed material sizes on gravel bars or along bars at the edge of the river (use a "by number" rather than "by weight" analysis). The material on the surface is important in scour calculations – the surficial material will "armor" or pave the riverbed thus limiting the scour potential. The finer sand to silt sizes commonly

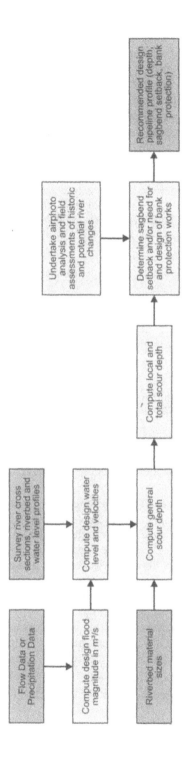

Figure 1 Buried River Crossings – Overview of Design Steps

78

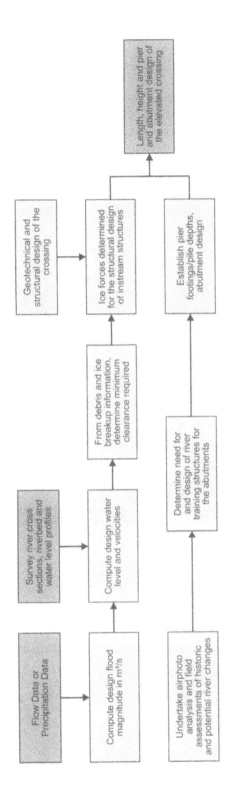

Figure 2 Elevated River Crossings – Overview of Design Steps

intermixed with gravel/cobbles/boulders is not key in computing scour depth. For sand bed rivers, containing few if any gravel or boulder sizes, undertake a grain size analysis of a bulk sample to establish bed material sizes.

- Compute general scour or general riverbed lowering due to the design flood, using various empirical methods or computer models.

- Estimate local scour that could occur at channel bends or at confluences of channels or near bridge piers or river training structures. From an analysis of extensive empirical data for gravel bed rivers (Hydrocon Engineering, 1983) and pipeline operational experience, local scour is generally much more significant than general scour.

- Estimate from field and airphoto assessments, the potential long term future bank erosion and channel changes that could affect the crossing. Determine whether to provide sufficient setback of the sagbend and thus pipe profile into the bank (or adequate length of elevated crossings) for the potential bank erosion or whether to armor the bank or bridge abutment, thus protecting the pipeline crossing. The optimum method depends on site-specific river conditions, pipeline bending considerations and economics. For economic, pipe bending and thermal reasons, excavation into a steep unstable slope may not be desirable thus necessitating bank armoring structures. The impact of construction on the stability of the riverbanks, particularly in permafrost areas, should be considered in establishing sagbend locations.

- Where the pipeline is parallel to, or in the main channel or floodplain, estimate the potential bank erosion, channel changes and the development of new channels. A channel change upstream caused by a landslide or bank erosion or accumulation of debris can change flow conditions near to or over the pipeline.

3.2.7 Field Data Requirements and Purposes

The field data requirements, addressing the how, where and why, are provided in Table 2. The table discusses:

- comparative airphotos;
- topographic plans;
- river and floodplain cross sections;
- river and water level profiles;
- highwater and ice marks;
- bed material sizes;
- bank material and characteristics;
- floodplain characteristics; and
- the presence and nature of nearby existing facilities and structures.

80

All the data listed in Table 2 may not be necessary to the same degree for Major or Intermediate River crossings. Site specific river conditions and the crossing mode (buried or elevated) may, as discussed previously, affect the nature of and extent of the data requirements. The Hydrotechnical Design Engineer will determine the data needs for each crossing.

3.2.8 Design Methodologies

The data requirements and design steps were outlined in the previous subsections. In this subsection, the need for and methodology for determining the necessary design input via field data is explained in greater detail.

3.2.8.1 Flow

Computing flows is required for:

- The design of the pipeline crossing, bank protection measures and river training structures (the latter if needed).
- Construction planning re: assessing the feasibility of various flow isolation techniques during construction.
- The hydraulic design of temporary or permanent access road or workpad bridges or culverts.

Table 2 - Field Data Requirements for the Detailed Design of River Crossings

Type of Data	How, where and why data is obtained, needed and utilized
1. **COMPARATIVE AERIAL PHOTOGRAPHY**	As obtained from governmental or private sources.
	Aerial photography coverage during major floods is particularly useful to assess extent and characteristics of flows in the floodplain.
	Obtain at least three (3) comparative photographs over the longest possible time frame. Photographs before and after major floods are most useful.
2. **TOPOGRAPHIC PLANS**	As obtained from government agencies or as prepared specifically for the project by the owner.
	Used for design information, determining right-of-way requirements, drainage and erosion control measures, and for general construction planning.
3. **RIVER AND FLOODPLAIN CROSS SECTIONS**	Minor or Typical crossings, require the following:
	• Survey of the pipeline centerline from top of valley slope to top of valley slope. Survey interval to be sufficient to accurately define the river or creek channel, bank height, slope and profile of the approach slopes.
	Intermediate crossings typically require survey cross sections, perpendicular to the river and floodplain, at the centerline. Additional sections upstream and downstream, at a distance upstream and downstream of the centerline equal to two to three times the river's width may be required.
	Major crossings require survey cross sections similar to that for the Intermediate crossings.

Type of Data	How, where and why data is obtained, needed and utilized
4. RIVER AND WATER LEVEL PROFILE	A water level and thalweg (deepest part of the channel) profile should be surveyed to connect up the cross sections.
	The water level profile will require a high degree of accuracy particularly for low gradient streams. Measure water levels in quiet backwater or slow moving areas as measurements in high velocity areas may not be very accurate. Measurements can usually be done from the bank. Survey of water level drops at rapids and beaver dams, etc. are important.
	In the summer, the deepest part of the channel is usually readily located by experienced surveyors. Aerial photographs should be used to estimate the location of the thalweg if the surveys are done in the winter.
	The relative thalweg depths from place to place along the river are important. The maximum depth in a straight river reach versus at a bend or just down-stream of an island where subchannels rejoin is important in establishing the local scour potential of the river. The location of the survey and particularly major scour areas, should be noted on aerial photographs or topographical plans.
5. HIGHWATER AND ICE MARKS	High water and ice marks, needed to confirm or establish design water level calculations, are obtained during surveys and field reconnaissance. Ice-affected levels are useful for determining the height of bank protection, the required height of elevated crossings, or the required elevations of adjacent pipeline appurtenances such as valves.
	High water marks can be obtained as follows:
	• During spring breakup, high water marks are readily identified in the snow.

Type of Data	How, where and why data is obtained, needed and utilized
	• Identifying silt and debris levels or marks left in the floodplain after flood events. Where ice has moved over a vegetated area pushing down the trees and bushes, high water marks on vegetation may not be accurate. Marks on large trees which do not bend, are accurate. High water marks on existing bridges, provided a high flood or breakup event has occurred during the life of the structure, are valuable.
	• High water levels can often be obtained from long-term residents in the area, especially those who experienced the last major flood.
6. BED MATERIAL SIZES	Bed material sizes are required to compute scour depths.
	First, a broad classification of the river bed material type should be made with respect to standard size gradation ranges for cobbles, gravel, sand, silt and clay.
	For cobble and gravel bed rivers, bed material size should be sampled at the upstream end of mid-channel bars, islands or point bars (in the inside of a bend) or side bars (in straight river reaches). For a sand bed river, sample the bottom material at the point of flow crossover (midway between channel bends) or near the upstream end of a bar. For silt and clay bed rivers, sample at the crossover location.
	For cobble and coarse gravel sizes, determine the average material size by:
	• Laying a grid pattern over the area.
	• Photographing the grid.
	• Measure the material sizes relative to the grid and determine the gradation by number. It has been shown by various researchers that plotting by number, provided at least 50 pieces are measured, is more representative than a gradation analysis by weight which can be significantly influenced

84

Type of Data	How, where and why data is obtained, needed and utilized
	by one or two large bed material sizes. If a grid is not available, obtain a photograph using a ruler, notebook or lens cap for scale reference.
	For fine gravel, sand, silt and clay rivers, obtain a representative bulk sample and undertake a sieve analysis in the laboratory.
	Boreholes can be used for size gradations if an undisturbed sample can be obtained and if the larger sizes are not excluded. They are not recommended to determine the bed material size of gravel or cobble bed rivers as the material extracted from the boreholes generally do not represent the larger sizes. Boreholes are useful to indicate whether significant material size variations exist vertically a factor that can affect the scour potential.
7. BANK MATERIAL AND CHARACTERISTICS	Wide variations generally occur in bank material. General conditions can be documented via photographs. Distinct changes in material, from sand to gravel, for example, should be documented.
	Vegetative cover is best documented via photographs.
	Natural bank erosion, upstream and downstream, is generally a good indicator of what could happen at the pipeline crossing. It should be determined if the bank erosion is local, at a bend or at an obstruction, or if it covers a broader area. Photographic documentation is the best technique.

85

Type of Data	How, where and why data is obtained, needed and utilized
8. FLOODPLAIN CHARACTERISTICS	On a crossing with wide floodplains, subject to frequent flooding, a thorough documentation of floodplain conditions by an experienced hydrotechnical engineer is essential. The pipeline design selected for the floodplain (deep burial or nominal cover or the use of river training structures) can have a greater impact on the construction cost and maintenance of the crossing than even the main channel section. Documenting the floodplain characteristics in the field and from air photos is primarily a qualitative assessment focusing on questions such as: • Have subchannels developed in the floodplain? Where are they located? Are they directly connected to the main channel at the upstream end? • What conditions upstream in the main channel, if any, could result in a dramatic increase in flow in the floodplain subchannels? Could a debris or ice jam cause new channels to form in the floodplain or cause existing channels to enlarge? • From comparative aerial photographs, have floodplain changes occurred elsewhere upstream and downstream? Is there a possibility that similar changes could occur at the proposed crossing?
9. ADJACENT EXISTING FACILITIES AND STRUCTURES	Major structures in the vicinity of the proposed crossing can be identified from aerial photographs or can be documented photographically in the field. If deemed necessary by the hydrotechnical engineer, additional data for the structures, that may be useful for the pipeline crossing design, can be obtained. For example, flood design water levels and icing levels at the adjacent highway bridges are useful as a check on the design water level for the pipeline crossing.

86

The design flow is computed from recorded flow data for the rivers to be crossed or from regional flow data in the area with similar hydrologic conditions. (It should be noted that in mountainous regions, hydrologic conditions may be dramatically different from one valley to the next). Lacking recorded flow data, the design flood can be computed from rainfall data and complex rainfall - runoff models provided regional rainfall and flow data for a high flow event is available to enable calibration of the model.

The importance of the magnitude of the design flow depends on the characteristics of the river crossing. Namely:

- if the crossing has a wide floodplain, the precise magnitude of the design flow is not very critical,
- if the crossing is narrow with high banks, the design flow and thus resultant water level is very important especially if an elevated crossing is utilized or river training structures are required.

For construction planning, typical rather than the flood design flows, are necessary. Potential ranges of flows for the construction period are generally adequate to determine the feasibility of various flow isolation techniques such as pumping or fluming.

3.2.8.2 Water Level

The design water level is required:

- to compute the main channel scour;
- to determine the depth and magnitude of floodplain flow which determines the potential for new channels to form and floodplain scour and thus the required pipeline depth in the floodplain; and
- to establish the height of river training structures, bank protection measures, elevated crossings and thus the required elevation of above ground valves.

The design water level is computed from the following information and analysis:

- River channel and floodplain cross sections. A minimum of three cross sections are required.
- Streambed and water level profiles as per the surveys. The surveyed elevation of highwater marks on the riverbanks or nearby structures, are used as a check on the computed water levels.
- Estimating channel and floodplain roughness values from bed material sizes or from known data for comparable river conditions. From recorded water level and flow measurements, the channel roughness values can be "back calculated" for the flow measurement location.

- Utilizing simple methodologies such as the Manning's Equation or one-dimensional computer programs such as the U.S. Army Corps of Engineers' HEC-RAS model.

In complex river systems such as deltas, a two-dimensional model may be required to predict water levels accurately. These models require extensive multi-year field data, to ensure that actual water level conditions are well represented by the model.

In ice-affected rivers the following winter and spring breakup conditions may affect the water level at a crossing:

- Flow over ground-fast ice (commonly known as aufeis or icings). Icings form in low flow rivers, especially in shallow wide braided watercourses. The thickness or level of the ice varies from year-to-year and from location to location depending on snowfall, temperature and magnitude of the base flow to the river. For many northern rivers, the initial spring breakup flow over the ice produces water levels often far exceeding the water level due to the design open water flood. The spatial and temporal variability of icing conditions precludes precise computational methods for site specific locations. Historic information, especially anecdotal from locals, is valuable in combination with an understanding the physical characteristics of the river and floodplain, which may define the upper limits of ice and thus spring breakup water levels.
- Thermal or frazil ice cover. A thermal and smooth ice cover forms in slow moving rivers. The water level for thermal ice conditions is relatively consistent from year-to-year and is best determined from field observations, survey and local knowledge. In high velocity streams, frazil ice forms to a thickness and level such that the backwater created by the accumulated ice reduces velocities to the point that the development of the ice cover proceeds upstream. Ice, especially a frazil ice cover, increases the overall roughness of the channel and therefore, compared to open water flow conditions, a winter water level is often higher than a summer water level for the same or lower flow magnitude. Ice scars on trees are often a good indicator of historic ice levels.
- Spring breakup. Rapid breakup on major rivers can produce extreme water levels and flows as jams form and release. Conditions are very site specific. On major northern rivers such as the Mackenzie and the Yukon, ice jams greater than 10 m in height occur. Historic ice jams are generally readily determined from field observations and from local knowledge.

Typical water levels and depths during the construction period are used to determine equipment operational limitations and to plan water and sediment management techniques.

3.2.8.3   Scour

Scour of a riverbed occurs in response to high flows and secondary currents at bends or channel confluences. Scour definitions are provided on Figure 3. General scour is the overall lowering of the riverbed in a straight uniform channel in response to the design flow. As flows increase, velocities increase and the ability of the river to transport or scour material from the bed increases.

Various techniques are available to compute general scour. They include regime methods (Blench 1969), competent velocity methods and complex sediment transport models.

In the regime method, Figure 4, a river is said to be in regime or in equilibrium if its width and depth are stable for a specific flow, slope, bed material size and sediment load. The competent velocity method provides relationship between velocity, bed material sizes and flow depth (Figures 5 and 6). As the velocity increases during a flood, the ability to scour and transport bed material increases and the flow depth increases.

Mathematical models such as the U.S. Army Corps of Engineers' HEC-6, initially developed to model depositional and erosional processes upstream and downstream from dams have been used by some designers. Extensive field data is required as input to the model. Variations of the Corps' model have been developed by others.

With respect to selecting the appropriate general scour model, it is the Author's strong view, from extensive annual reconnaissances, surveys and operational experience, that:

- The general scour is generally small compared to the local scour at bends or confluences and therefore the general scour method selected is not very important in establishing the required pipeline profile of buried crossings.
- Some designers or regulatory jurisdictions favour the regime method while others primarily use the competent velocity method or similar techniques. Past experience with a particular method rather that the validity of the method generally determines which approach is favoured. Models such as HEC-6 require significant field data, which is often not available.
- The regime method is a sound approach to determine if the stream is in equilibrium. For example, for a pipeline crossing of an alluvial fan, it provides a check or whether the fan is in a depositional or erosional mode

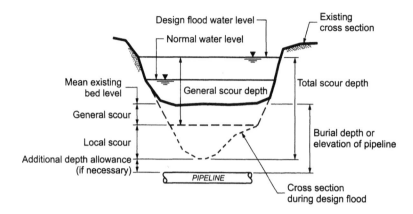

1. **General scour** is the general lowering of the channel bed that occurs in a straight uniform channel during the design flow event. General scour is measured from the **mean existing bed level in a straight channel section**.

2. **General scour depth** is the mean depth of the channel bottom below the **design flood water level.**

3. **Local scour** is site-specific scouring that occurs due to secondary currents at channel bends, constrictions, piers or obstructions to flow. Local scour is measured from the bottom of the general scour elevation.

4. **Total scour depth** is the difference between the **design flood water level** and the local scour elevation.

5. **Burial depth** of the pipeline is the distance from the existing bed level to the top of the pipeline. The burial depth is equal to the company's minimum cover depth criteria or computed scour depth whichever is greater. An **additional allowance** may be made for uncertainties regarding potential scour.

6. For **major crossings** with a depth greater than the minimum depth criteria, the **pipeline burial depth should be specified as an elevation** rather than as a depth since the existing bed level is variable across the width of the channel and may change from the design to the construction phase. For **minor crossings** buried to the minimum cover depth criteria, the **pipeline burial depth may be specified as a depth of cover**.

Figure 3 Definitions for Buried Crossings

Mean or General Depth Determination (Blench, 1969)*

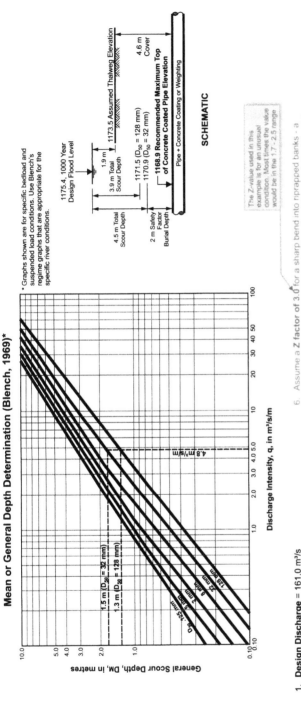

Discharge Intensity, q, in m³/s/m

General Scour Depth, Dm, in metres

4.8 m³/s/m

1.5 m (D₅₀ = 32 mm)

1.3 m (D₅₀ = 128 mm)

Dm 185 mm

8 mm

22 mm

128 mm

SCHEMATIC

1175.4, 1000 Year Design Flood Level

1173.5 Assumed Thalweg Elevation

1.9 m

3.9 m Total Scour Depth

4.5 m Total Scour Depth

1171.5 (D₅₀ = 128 mm)

1170.9 (D₅₀ = 32 mm)

1168.9 Recommended Maximum Top of Concrete Coated Pipe Elevation

4.6 m Cover

2 m Safety Factor Burial Depth

Pipe + Concrete Coating or Weighting

The Z-value used in this example is for an unusual condition. Most times the value would be in the 1.7 - 2.5 range

A safety factor is recommended only for crossings with a high level of uncertainty. It is not a normal requirement, as the appropriate degree of design conservatism for the project will be considered in establishing the Z-factor.

1. **Design Discharge** = 161.0 m³/s

2. **Design River Width** = 50.0 m

3. **Discharge Intensity** = 161.0 / 50 = 3.22 m³/s/m. Increase discharge intensity by 50% to provide for non-uniform flow distribution across the 50 m width = 1.5 x 3.22 = 4.8 m³/s/m.

4. **Bed Material Size** = average size variable as observed from site conditions and photos. Estimate a range from D_{50} = 32 mm to D_{50} = 128 mm.

5. **General Scour** = 1.5 m (for D_{50} = 32 mm) and 1.3 m (for D_{50} = 128 mm).

6. Assume a **Z factor** of 3.0 for a sharp bend into riprapped banks - a conservative value as per Fig 3.8. Thus Total Scour = 3.0 x 1.5 = 4.5 m (for D_{50} = 32 mm). 3.0 x 1.3 = 3.9 m (for D_{50} = 128 mm).

7. Determine **Net Scour Depth** (below the riverbed) for the total flow depth of 4.5 m or 3.9 m.
   - Design Depth of flow (assuming no scour) = 1.9 m.
   - Riverbed thalweg elevation = 1173.5
   - Design flood water level = 1173.5 + 1.9 = 1175.40 m
   - Thus scour elevation = 1175.4 - 4.5 = 1170.90 for D_{50} = 32 mm and 1175.4 - 3.9 = 1171.5 for D_{50} = 128 mm

8. Recommended additional **Safety Factor** = 2 m
   - Thus 1170.9 - 2.0 = 1168.9 for D_{50} = 32 mm and 1171.5 - 2.0 = 1169.5 for D_{50} = 128 mm.

9. Recommended Maximum Top of Concrete Coated or Concrete Weighted Pipe Elevation Between the Sagbends = 1168.9 m

Figure 4 Regime Method Scour Computation

Neill (1973, p.90) recommends an iterative graphical procedure for determining general scour in straight uniform channels with no bed load. A *hypothetical example* for a pipeline crossing is given here.

Design flow	500 m³/s			
Bed material size	30 mm	(or 0.1 ft)	median size as determined by number

Calculations begin with Iteration 1, which is a trapezoidal approximation of the surveyed river cross section. The assumed values for the cross section are not realistic - the velocities are too high for the bed material size - but are used only to illustrate the calculation methodology.

	Iteration 1		Iteration 2		Iteration 3	
	SI units	Imperial units	SI units	Imperial units	SI units	Imperial units
Depth	3.5 m	11.5 ft	5.0 m	16.4 ft	5.75 m	18.9 ft
Top Width	40.0 m	131 ft	40.0 m	131 ft	40.0 m	131 ft
Bottom Width	24.0 m	78.7 ft	24.0 m	78.7 ft	24.0 m	78.7 ft
Flow Area	112.0 m²	1206 ft²	160.0 m²	1722 ft²	184.0 m²	1981 ft²
Mean Velocity	4.5 m/s	14.6 ft/s	3.1 m/s	10.3 ft/s	2.7 m/s	8.9 ft/s
Crtical Velocity		8.2 ft/s		8.6 ft/s		8.9 ft/s
Comment	Assume a depth of 11.5 ft (3.5 m). Mean Velocity (14.6 ft/s) is greater than Critical Velocity (8.2 ft/s) therefore scour occurs. Go to a deeper cross section in Iteration 2.		Assume a depth of 16.4 feet (5.0 m). Mean Velocity (10.3 ft/s) is still greater than Critical Velocity (8.6 ft/s) therefore scour occurs. Go to a deeper cross section in Iteration 3.		Assume a depth of 18.9 ft (5.75 m). Mean Velocity (8.9 ft/s) equals Critical Velocity and therefore the flow depth is in equilibrium (no scour or deposition occurs). Calculation completed.	

Refer to Figure 6 for the graphic results.

· The choice between using the Regime Method (Figure 4) and the Method on this Figure should be made in the detailed design phase. Local scour must be added to the results of this method (see Figure 8)

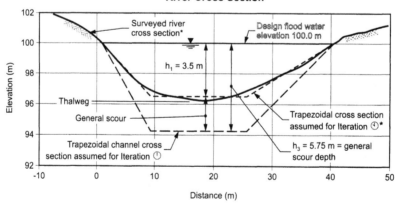

River Cross Section

* Use only cross section(s) in a straight river reach. A cross section at a bend would already represent local scour conditions and thus should not be used for determining the magnitude of general scour.

Figure 5 Competent Velocity Method Scour Calculations

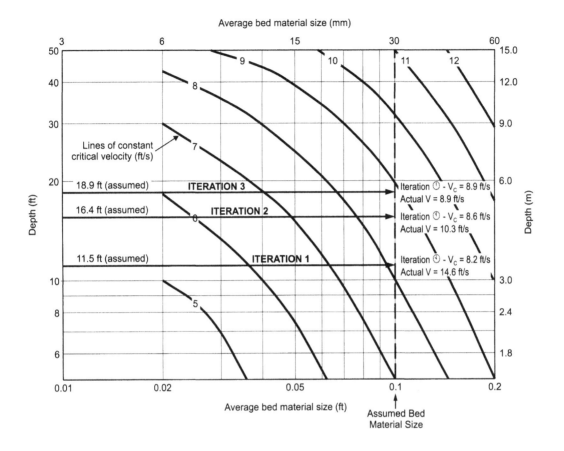

Critical velocities for significant bed movement of cohesionless materials in terms of bed material size and depth of flow. (After Neill, C.R., (editor) _"Guide to Bridge Hydraulics"_, 1973, Figure 4-12.)

In the final Iteration ①, the computed mean channel velocity (8.9 ft/s) is equal to the critical velocity (8.9 ft/s). The difference between the bed elevation from the survey (assumed to be Iteration 1) and Iteration 3 (2.8 m) is the general scour. The general scour does not account for local scour which may occur due to channel bends or obstructions to flow.

Refer to Figure 5 for the calculations.

Figure 6 Graphic Representation of Competent Velocity Method

– the influence of the receiving stream can significantly affect the slope and thus scour on the fan. The regime method can also be used to indicate the presence and influence of downstream controls such as bedrock outcroppings.

- The selected bed material size is a key component of all general scour methods. Figure 7 illustrates the effect of bed material size on flow or scour depth. As the average size increases, the scour depth decreases. For gravel bed rivers, the surficial bed material or the "armor" layer established by the river, should be used and not the average size from a bulk sample taken from a test pit at depth. The rationale is that it is the surficial sizes that determine the scour depth and not the finer interspersed material sizes.

Flow concentrations and secondary currents at channel bends and confluences or next to instream structures such as bridge piers generate local scour. From extensive empirical data on gravel bed rivers, the relationship between local and general scour has been established as indicated on Figure 8 (Hydrocon Engineering, 1983). Local scour is generally several times greater than general scour. A non-eroding rock or armored channel bend produces higher local scour than a free-eroding bend. A sharp-angled confluence of two relatively equally sized channels produces high local scour.

In determining the potential local scour for the design flood, the designer must consider the existing bends and channel confluences as well as reasonably foreseeable future flow patterns at the crossing. Determining local scour potential is thus dependent on a qualitative rather than quantitative assessment. The experience of the designer is the key in producing reasonable yet prudent total scour (general scour times the local scour multiplier) which determines the required top of pipe elevation. Scour multipliers ranging from 1.5 to 3.0 for straight channels and sharp channel confluences respectively can be experienced. In other words for a 1.0 m deep straight channel, the depth of flow at a minor bend and sharp confluence may be 1.5 m and 3.0 m, respectively.

The computed scour determines the required top of pipe profile. In special cases, an additional safety factor might be added. In most instances, the design should be reasonably conservative and will not require an additional safety factor.

3.2.8.4   Bank Erosion

The potential bank erosion determines the required pipeline sagbend setback, the sagbend to overbend pipe profile and the need for bank protection measures if the sagbend cannot be located sufficiently far enough into the bank for economic, technical, and environmental or construction reasons.

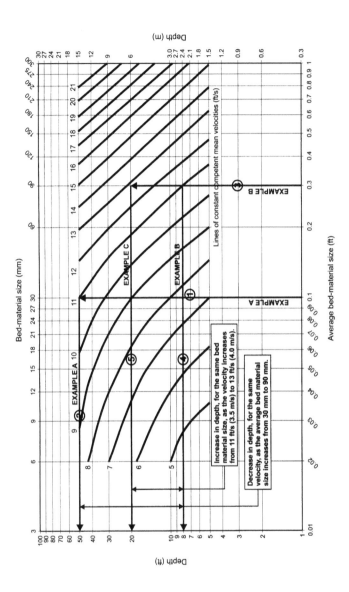

Figure 7 Effect of Bed Material Size on General Scour Depth

NOTES:

1. EXAMPLES ILLUSTRATE THE :
   - effect of bed material size on depth (for a given velocity),
   - effect of depth on critical velocity for a given bed material size

2. EXAMPLE A
   For a bed material size of 0.1 feet (30 mm), ⓐ a flow depth of 50 feet (15.2 m) ⓒ will result from a velocity of 11 ft/s (3.5 m/s).

3. EXAMPLE B
   For a bed material size of 0.3 feet (90 mm), ⓒ a flow depth of only 8 feet (2.4 m) ⓓ will result from the same velocity of 11 ft/s (3.5 m/s).

4. EXAMPLE C
   For the same bed material size of 0.3 feet (90 mm) ⓒ, the flow depth will be increased to 20 feet (6.1 m) ⓔ if the velocity is increased from 11 ft/s to 13 ft/s (3.4-4.0 m/s).

95

Z-Factors for Free-Eroding Bends
(as per research by Hydrocon Engineering, 1983)

- ▲ N. Saskatchewan River (Doyle et. al., 1979)
- ⊡ N. Saskatchewan River (Nwachukwu et. al., 1972)
- ✕ TAPS Alaska data
- ⊚ Athabasca River (Neill, 1973)
- ★ Mendenhall River at Juneau, Alaska (Hydroconsult, 1997, see Photo below)

Example of Z-Factor Determined by Field Measurements
Mendenhall River at Juneau, Alaska

A 1:50-year flow (530 m³/s) in September 1995 caused severe scour in the area shown at a riprapped bend. Post-flood surveys indicated a Z-factor of 2.3. Maximum scour depths, during the peak of the flood were likely greater. Local scour depth is greater next to an armored or rock bank than next to a free eroding bank.

Figure 8 Local Scour Multiplication Factors

96

Bank erosion potential is determined from:

- An assessment of historic erosion as determined from comparative aerial photographs and local information. Historic air photos are especially valuable if pre and post major flood photos are available – the vast majority of bank erosion occurs during high flows (Veldman 2002),
- An assessment of the potential for a bend and thus increased bank erosion to occur at the pipeline crossing location,
- An assessment of the timing of major flow events. In northern climates, high flows in spring breakup, when the banks are frozen, produce minimal bank erosion whereas a late summer flood generally produces maximum erosion (Veldman, 2002),
- An assessment of the impact of construction on the stability and erosion of the riverbanks. Construction disturbance, especially at a steep, high and frozen bank, can lead to local increased bank erosion and thermal degradation.

3.2.8.5   Channel Changes

Under high flow conditions, new channels may form in floodplains or subchannels may become enlarged to form main channels. The potential for this occurring is best determined from field assessments by experienced river engineers and from a review of comparative airphotos. With respect to addressing the impact of the formation of or enlargement of floodplain channels on the pipeline profile, the options are:

- If it could likely occur, a horizontal deep buried pipe profile extending the full width of the main channel plus floodplain is appropriate,
- If the development of a new main channel is unlikely in the floodplain, the burial depth specified for the floodplain can be less than that for the main channel.

The final pipe profile depends in part on the proximity of the main channel to and the width to the floodplain. If the main channel and floodplain are close and the floodplain is not extensive, a horizontal pipe profile would likely be specified. If on the other hand the main channel and floodplain are some distance apart and the floodplain is wide, a variable pipe elevation may be appropriate and sound. (Depending on the pipe diameter, this may require an intermediate sagbend in the floodplain or alternatively the pipeline can be "roped in" from the floodplain level down to the main channel depth of burial.)

3.2.8.6   Design Examples

Figures 9, 10 and 11 illustrate example design drawings for Major, Intermediate and Typical buried crossings respectively. Figures 12 and 13 detail the design steps, assessment and recommendations for a major crossing.

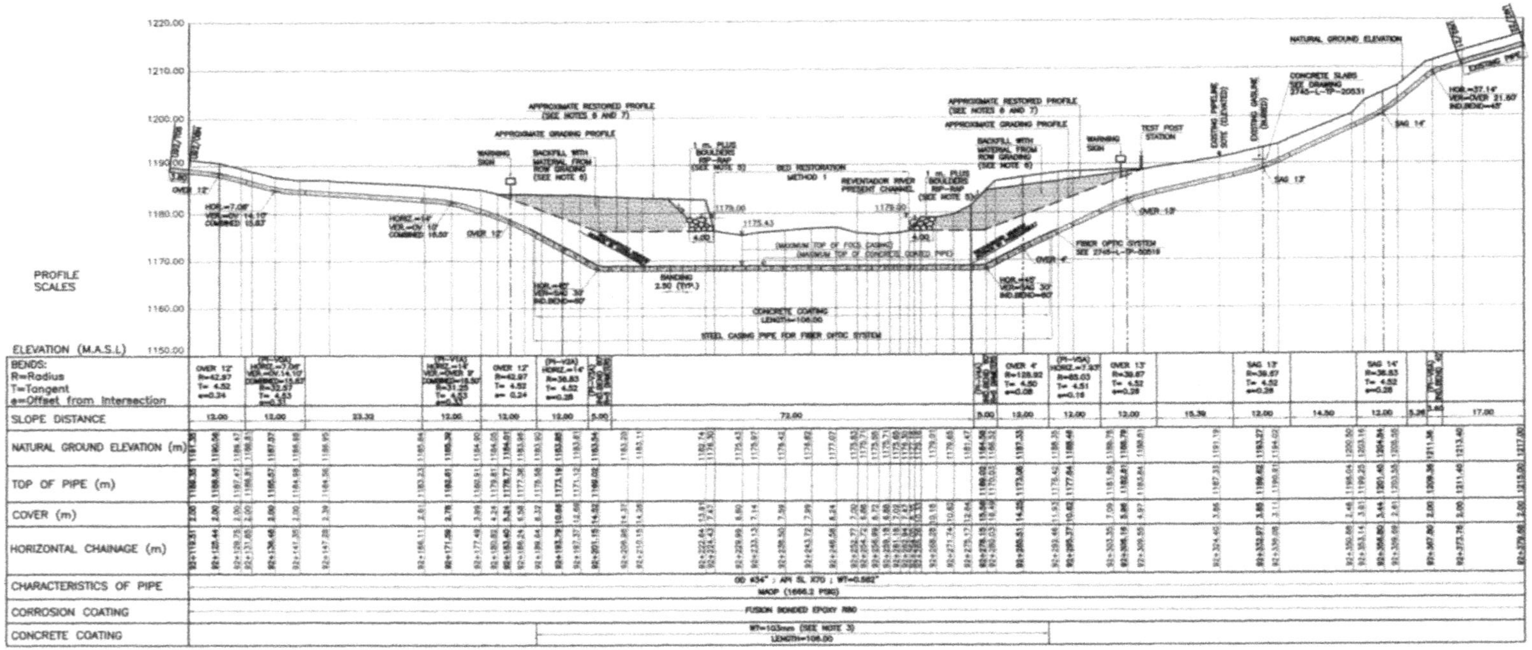

- A representative detailed design profile for a major river crossing. The pipeline elevation, sagbend location and pipe profile into the banks are as per the river engineering recommendations. Pipe bending considerations determine the precise profile of the pipeline.
- Note that the top of pipe elevation is specified by an elevation rather than cover depth from the thalweg.

Figure 9 Example, Major Buried River Crossing

INTERMEDIATE RIVER CROSSING

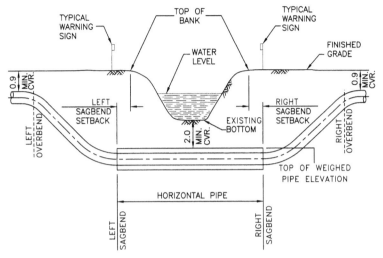

KP	CROSSING	TOP OF WEIGHED PIPE ELEVATION	LEFT SIDE		RIGHT SIDE	
			SAGBEND STATION	OVERBEND STATION	SAGBEND STATION	OVERBEND STATION
		A detailed table of dimensions, as developed from the river engineering and pipeline design, defines the specifics for each Intermediate Crossing. Thus a separate Detailed Design Drawing is not required.				

- Top of pipe is normally specified by elevation rather than cover depth from thalweg. The 2 m cover shown is a typical minimum depth criteria.
- Site specific analysis required for detailed design. (ie. cover depth, sagbend setback and overbend location).
- Note that in most cases both the sagbend and overbend stationing need to be specified in order ro ensure adequate setback for the pipeline into the riverbanks.

Figure 10 Example, Intermediate River Crossing

99

TYPICAL RIVER CROSSING

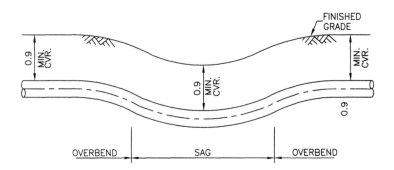

Figure 11 Example, Typical River Crossing

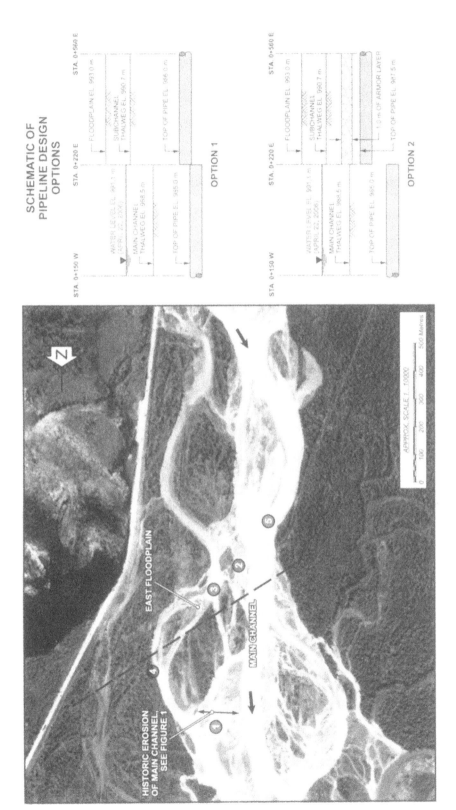

For Design Notes, See Figure 13

Figure 12 Airphoto and Design Options for a Major River Crossing. (Canada)

101

1.0 DETAILED ASSESSMENT

1.1 The main channel of the River has been along or near the west bank since 1978. Just downstream from the proposed alignment ⓐ, the main channel has:

- Been located midchannel in 1978,

- Swung sharply eastward by 1997 as it eroded through a mid-channel vegetated area,

- Then by 2005, the channel straightened again and was located in midchannel.

1.2 In view of the mobility of the main channel of the River in this reach, a deflection of the main channel into the eastern subchannels ⓐ is a strong potential. At the present the magnitude of flow in these subchannels is reduced by debris jams at their mouths - at a high enough flow, the debris could float out and open up the channel.

1.3 Thus the **pipeline across the main channel and eastern subchannels should be designed for**:

- Main channel scour. Maximum scour in a multi-channelled river occurs at channel confluences. At location ⓐ, this is evident - a 1.0-2.0 m deep hole exists where two relatively small subchannels come together,

- Erosion of the east bank of the eastern subchannel ⓑ.

1.4 The maximum **surveyed scour** in the main channel occurs at the point where the riprap projects out ⓒ. The September 16, 2005 review indicates an existing maximum scour elevation of about 983.5 m compared to 988.0 - 988.5 in the area of the existing crossing. As a comparison at the proposed crossing, the surveyed thalweg and design top of pipe elevation are 988.5 m and 985.0 m respectively.

1.5 The **deep local scour at 983.5 m**, 1.5 m deeper than the proposed main channel top of pipe elevation at the proposed line, is a local phenomenon due to the presence and shape of the west bank riprap. It **is thus not a design basis for the proposed crossing** of the main channel or subchannels.

1.6 The potential scour depth in the east subchannel is computed as follows:

- Top of east bank = El. 993.0,

- Present depth of subchannel = 993.0 - 990.6 = 2.4 m,

- General deepening of subchannel due to increased flow if more flow or if a main sub-channel is deflected into the east channels = 0.5 m,

- total depth of subchannel if a flow confluence forms in the subchannel = (2.4 + 0.5) X 2.5 (local scour multiplier) = 7.25 m,

- Scour elevation 993.0 - 7.25 = 985.75 m say 986.0 m.

1.7 The magnitude of eastward bank erosion of the main channel downstream of the crossing between 1978 and 1997 is about 150 m. In a 6 year or less period from 1978 - 1984, this same main channel eroded eastward about 75 m through a low and partially vegetated floodplain.

1.8 The east bank of the easterly subchannel at Sta. 0+495 is well defined and well vegetated with mature trees, see site photos. Thus the potential for erosion of the east bank will be less than that of the lower and less vegetated floodplains.

2.0 CONCLUSIONS

2.1 The computed **scour depth** for the east subchannel / floodplain area is 986.0 m or 1.0 m higher than the present design elevation for the main channel.

2.2 A **main channel top-of-pipe design elevation** of 985.0 is recommended from the west sagbend at Sta. 0+150 W to Sta. 0+220 E.

2.3 Scour and bank erosion of the **east floodplain** results in a design value of 986.0 m from Sta. 0+220 E to Sta. 0+560 E which represents a sagbend setback of 60 m from the edge of the easterly subchannel.

3.0 RECOMMENDATIONS

3.1 Design top of pipe (concrete weighted) elevation of 985.0 m from Sta. 0+150 W to Sta. 0+220 E.

3.2 For the east floodplain, a design top of pipe elevation of:

- 986.0 m from Sta. 0+220 E to Sta. 0+560 E or,

- 987.5 m within the same stationing if an armor layer is placed above the pipe in the ditch line.

For Airphoto, see Figure 12

Figure 13 Assessment and Recommendations for a Major River Crossing. (Canada)

3.2.8.7 Unique Design Conditions and Solutions

Alluvial Fans

Alluvial fans formed at the mouth or delta of mountainous streams pose a unique design challenge. The fans are characterized by deposition, rather than scour however the main channel can suddenly switch to and form at any location on the fan. New channels can form in previously vegetated areas.

Due to the cone-shaped profile of the fan, a horizontal pipe profile is generally not practical but rather a minimum burial depth criteria, relative to the main channel, is generally used. For example for a surveyed main channel depth and minimum pipeline cover depth criteria of 1.0 m and 2.0 m respectively, the pipe profile would provide 2.0 m and 3.0 m of cover at the main channel and for the remainder of the fan respectively.

Glacier Dammed Lake Releases

If the watershed contains glaciers, glacier dammed lakes can form at tributaries blocked by the glacier and suddenly release in a regular or less regular intervals. Snow melt and rainfall can result in rapid filling of the lake which, due to the resultant head, suddenly forms a drainage route or tunnel through the glacier. At times, the release can occur in the middle of winter leading to ice breakup and jams.

Local knowledge is key to assessing the potential for and impact of lake releases on flow and water levels at the crossing. Figure 14 illustrates an Alaskan glacier and the impact of lake releases on annual peak flows.

Debris Flows

During major natural events in the watershed such as landslides or volcanic eruptions combined with or without heavy rainfall, extremely high debris or sediment loads can be transported by the stream. The event is generally characterized by deposition, high sediment and water levels, and rapid and frequent channel changes. Following the event, new channels will be scoured through the deposited material.

Local knowledge of historic conditions is important in assessing the potential impact of debris flow on the depth of burial and often more importantly, on the design width of the crossing. Establishing a burial depth below the pre-event surveyed level is important as the post event deposition level is generally not a stable condition. Knowledge of the magnitude of historic debris flow events is an important input into assessing the relative merits of a buried or elevated crossing. An example of pre and post volcanic eruption conditions for a mountainous stream is illustrated on Figure 15 along with the required construction to ensure a pipeline profile into the stable pre-event grade. Figure 16 illustrates the failure of a buried gas pipeline river crossing in a rock canyon due to debris flow.

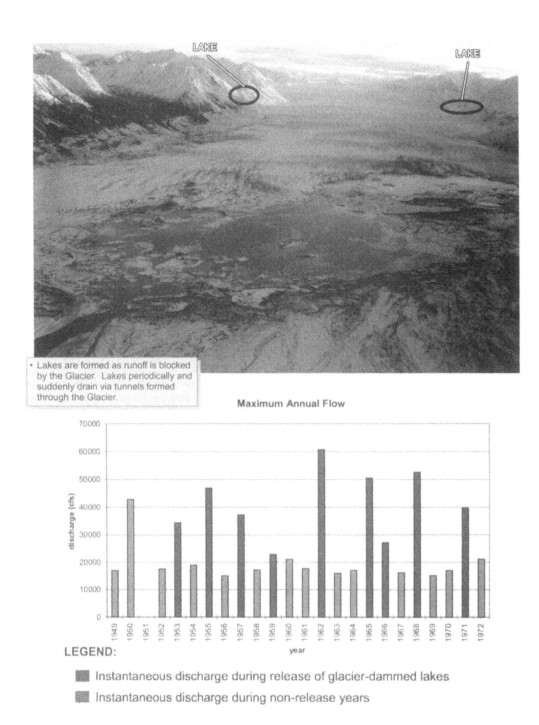

Figure 14 Example, Impact of Glacier Dammed Lakes on Flow (Alaska)

a) Rio Montana, Ecuador before volcanic eruption in 2002.

b) Eruption of Mount Reventador.

c) Rio Montana after the eruption.

d) Main Highway Bridge inundated by debris flow.

e) Depth of excavation, approximately 7 m, required to install the pipeline into the stable pre-volcanic eruption streambed profile.

Figure 15 Example, Debris Flow (Ecuador)

a) Debris flow resulted in the complete removal of the pipeline crossing. The depth of the debris flow in the canyon was 10 - 15 m.

b) Close-up of the rock canyon at and upstream from the pipeline crossing. The rock canyon alignment was selected because of its non-erodibility. In rock areas, whether in the streambed or in the banks, the key to the integrity of the pipeline is the stability of the backfill.

c) Overall view looking upstream towards the canyon. The reconstructed line was realigned to the area of the path crossing the creek in the foreground, a zone of deposition downstream of the canyon. The total burial depth was about 7 m to ensure 3 m of burial into the pre-debris deposition stream grade.

Figure 16 Example, Debris Flow Causing a Complete Failure of a Gas Pipeline (Argentina)

Instream and Parallel Pipeline Alignment

In certain instances, the optimum pipeline alignment is instream in the main channel or in the floodplain or parallel to and adjacent to the river. Some of the reasons for these alignment scenarios may be:

- Technical - For hot oil pipelines in a permafrost environment, gravel bed rivers and floodplains generally provide a thaw-stable environment permitting burial rather than requiring a more expensive elevated line (Trans Alaska oil pipeline for example).
- Topographic - Steep or unstable slopes adjacent to the river may require sections of instream pipeline alignment. (Oil line in Ecuador and gas lines in Peru and Argentina for example).
- Land use - Intensive land use adjacent to the river, (city, town or farming) may require instream alignment. (Gas pipeline in Argentina which is located instream for more than 50 km).

Flow conditions and environmental requirements must permit instream construction. In Alaska, Argentina and Peru, winter or dry season flow conditions readily enable instream construction. In steep mountainous and seasonal streams, limited aquatic resources may produce minimal environmental impact.

In mountainous valleys, the optimum pipeline alignment is often parallel to and proximate to rivers and may require, initially at the time of construction and during the operation of the pipeline, river training structures to protect the pipeline. These structures could either be located on river bends where the pipeline is close to the river or involve a series of structures to protect the pipeline located along the fringes of the river.

Examples of instream and parallel pipeline alignments are presented on Figure 17. The input of experienced river engineers is required to ensure a sound and practical design. The potential need for and the method of operating and maintaining the pipeline and associated river training structures should be considered by the river engineer and owner in selecting the pipeline design in these types of unique conditions.

River Training Structures

River training structures may be used to:

- armor the banks and thus protect the sagbend- to overbend section of the pipeline;
- protect buried or elevated pipeline sections located parallel to or proximate to the river;
- guide flow through elevated crossings to protect the bridge structure;
- protect valve installations; and
- protect buried instream pipeline sections.

a) Example of deep buried instream alignment, (gas pipeline in Argentina). The instream alignment, selected due to intensive infrastructure in this valley in the Andes, was facilitated by a 6 month dry season.

b) Gabion sill protecting instream pipeline alignment (Argentina). Sill provides additional scour protection and protects the line from gravel mining operations in the river.

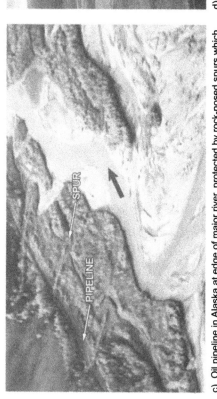

c) Oil pipeline in Alaska at edge of major river, protected by rock-nosed spurs which deflect the main channel velocities away from the shallow buried line.

d) Another example of spurs protecting an oil pipeline in Alaska.

Figure 17 Examples, Instream and Parallel Pipeline Alignments

The design of the structures, generally more so than that of the buried pipeline crossing itself, requires an accurate determination of design flow, water level (open water and ice-affected), velocities and scour (general and local). The input of a river engineer with considerable experience in design, construction methodologies and maintenance requirements is essential in the selection and design of the optimum river training structure. Some examples of structures are provided on Figure 18.

Frequently bank restoration measures, beyond typical and natural revegative growth, is required especially at high velocity river crossings with steep banks. Often, especially at rivers and adjacent rights-of-way which have numerous large boulders/rocks - pipeline backfill specifications generally preclude the placement of the large boulders close to the pipelines thus resulting in a surplus of these boulders - the naturally available boulders can provide, at little or no extra cost, adequate protection of the banks. Establishing various alternative restoration techniques, in conjunction with the construction supervisors, is recommended. (Also see the Construction Section 3.3).

Elevated Crossings

Elevated crossing, rather than a conventional buried crossing, may be optimal for technical, environmental, economic or construction reasons as follows:

- Technical - At thaw - unstable banks and river crossings in a permafrost environment, an elevated crossing may be required for a hot oil pipeline (Trans Alaska oil pipeline for example).
- Environmental - To avoid instream work, if a trenchless technique such as horizontal directional drilling (HDD) is not feasible, an elevated crossing may be necessary. For crossings of deep narrow gorges, an elevated crossing may be preferred over extensive grading required for a buried pipeline.
- Economic - As in the river condition cited above, an elevated crossing of a narrow gorge may be the optimum economic solution.
- Construction - If HDD or open cut is not feasible or practical at a deep high velocity river, an elevated single span crossing may facilitate construction and reduce schedule risks.

As in the case of river training structures, an elevated crossing requires an extensive evaluation of the design flow, water level, velocities and potential river changes. Debris flows, trees or bedload need to be considered in establishing the required clearance of the crossing.

The selection of the elevated versus buried mode and the optimum type of elevated crossing requires an evaluation by the river engineer and bridge design, construction and operations specialists. Some examples of elevated crossings are

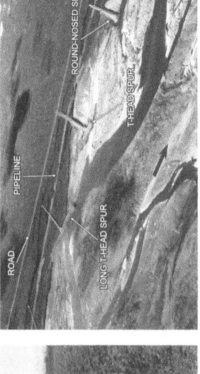

a) Riprap revetment.

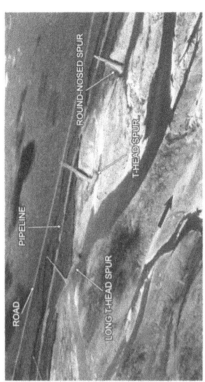

b) Spurs protecting buried pipeline. (Alaska)

c) Revetment / guidebank protecting a major pipeline bridge. (Alaska)

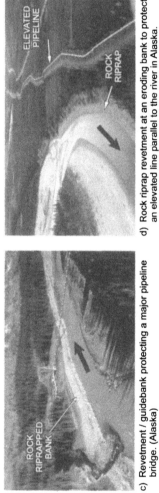

d) Rock riprap revetment at an eroding bank to protect an elevated line parallel to the river in Alaska.

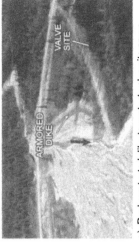

e) Rock protected dike to protect valve site an an alluvial fan. (Alaska)

Figure 18 Examples, River Training Structures

provided on Figures 19 to 21. A design example of an elevated crossing is provided on Figure 22.

3.3 Construction

3.3.1 Overview

Construction methodologies are dependent on schedule and technical, environmental and economic considerations. For example:

- Severe icing (aufeis) conditions in northern rivers may preclude the usage of flow isolation methods such as Aquadams, which require extensive removal of the ice for their installation. Very thick ice conditions may require pumping rather than fluming as the at-grade flume would require extensive removal of the ice downstream.
- The flow at the time of construction. The feasibility of achieving flow isolation via pumping and/or fluming depends on the flow. In severe winter conditions, the efficiency of both pumps and flumes can be dramatically decreased due to ice buildup necessitating either more pumps, larger or more flumes or a reduction in the number of streams that can be flow isolated.
- The feasibility of directionally drilled crossings is dependent on geotechnical and geological conditions under the streambed and in the approach slopes. The depth and width of the river valley may also affect the feasibility of a drilled crossing as well as the overall length of the drill.
- Environmental conditions, particularly at the time of construction, may dictate a particular crossing methodology.

The potential construction techniques for buried crossings are open cut with and without flow isolation. (The construction of elevated river crossings – see design example in Section 3.2 – are not discussed herein.).

The various construction techniques are illustrated by the following examples:

- Figure 23 illustrates the open cut technique used for a large 800 m wide crossing in North America. Large backhoes were employed for the deep excavation in the floodplain while backhoes and finally a Sauerman dragline were used for the main channel portion. The entire concrete coated crossing was pulled in.
- Figure 24 illustrates the open cut technique used for an intermediate crossing in South America. Water depths enabled backhoe excavation and "walking in" of the concrete coated pipe.
- Figure 25 illustrates a 500 m wide braided river in South America where dikes and diversions were used to excavate and lay the oil line in the

111

1. A **simple free span** of a narrow coulee with stable banks/slopes. Allowable length is a function of pipeline characteristics. Both ends could be buried. The elevated mode here eliminated the need for extensive grading of the right-of-way. (Alaska)

2. **Pile supports** across an alluvial fan. In the example shown, the normal overland piles are adequate for the imposed flow and ice conditions. The piles are designed for the computed scour depth. (Alaska)

3. Another view of a **pile supported crossing**. (Alaska)

Figure 19 Examples, Short Span Elevated Crossings

1. Single span **girder bridge**. Example shown is 180' long (54.9 m). (Alaska)

5. Close up of **girder bridge**, a simple box structure. Some riprap required on the north bank to protect the main north pier. (Alaska)

3. Another example of a single span **girder bridge**. Installation and maintenance of the pipeline on the bridge and the bridge structure are considerations re: selecting the buried or elevated mode. (Alaska)

Figure 20 Examples, Girder Bridge Crossings

113

1. An elevated crossing was used for this relatively small stream because of steep banks and uncertainties about scour depths in a very steep stream (Ecuador)

2. A 200 m long **suspension bridge**. In this example, uncertainties about the design flood and scour were some of the reasons why an elevated crossing was used. (Alaska)

High aquatic values in the river were a key reason for the elevated crossing, shown on Photo 3. With greatly improved horizontal drilling technologies now available, compared to the mid-seventies when the bridge was built, this river crossing would most likely be drilled at the present.

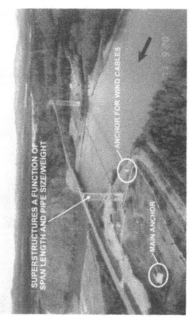

SUPERSTRUCTURES A FUNCTION OF SPAN LENGTH AND PIPE SIZE/WEIGHT

ANCHOR FOR WIND CABLES

MAIN ANCHOR

3. **A 370 m long suspension bridge.** Note expansion loops prior to the bridge to reduce stress on the bridge. Maintenance of the pipe/bridge requires specialized equipment/personel. (Alaska)

Figure 21 Examples, Suspension Bridge Crossings

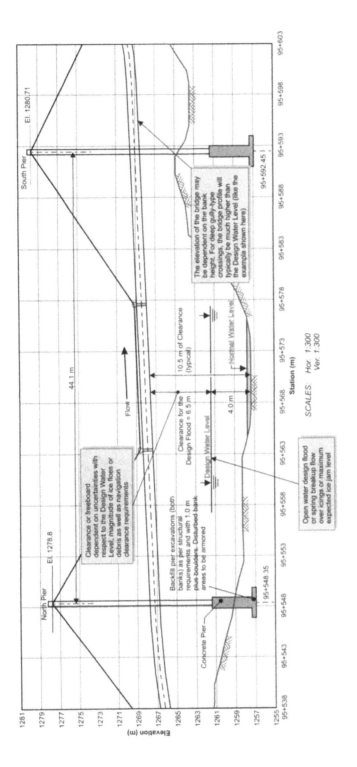

Figure 22 Design Example of Elevated Crossing (Ecuador)

115

a) Overall view of main channel (on the left) and wide floodplain (on the right). Total length of the crossing was about 800 m. (Canada)

b) Close up of main channel area. A Sauerman dragline was used to achieve the final grade in the main channel section.

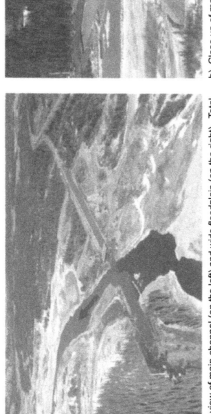

c) Concrete coated pipe on east floodplain ready for the pull in.

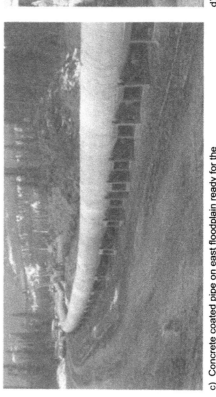

d) Line being pulled into the crossing.

Figure 23 Example, Open Cut Construction of a Major Crossing. (Canada)

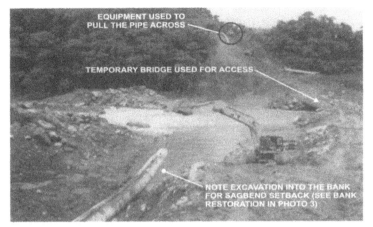

1. Concrete coated pipe about to be "walked" in by backhoe and booms (see Photo 2).

2. Water depth at time of construction determines whether pipe can be "walked" in as shown or needs to pulled in for a deep crossing.

Figure 24 Examples, Open Cut Construction of an Intermediate Crossing. (Ecuador)

117

1. Main channel diverted into the right side of a wide braided river while excavation is underway in the left side, in the foreground.

2. Another view from the right bank illustrating the left floodplain construction in non-flowing conditions. Flow magnitude too high to permit flow isolation.

3. Restored and completed river crossing.

Figure 25 Examples, Open Cut Construction with Flow Diversion. (Ecuador)

118

crossing in two steps – the diversions enabled the work to be undertaken in non-flowing conditions. Periodic high flows, during construction, required significant diversion dikes.

- Figure 26 illustrates open cut crossing conditions in an arctic climate. If there is no surface or near surface flow, the river crossing excavation may in fact be nearly in the dry although in one of the examples shown, minor seepage into the ditch created icings at the bottom of the ditch, a thawing and settlement concern for a hot-oil pipeline.

- Figure 27 illustrates flumes, culvert and pump isolation techniques used for several crossings in western Canada. With double flumes and culverts, the bypassed flow was 10 m^3/s believed to be the highest flow that had been bypassed up to that point in time. Excavation considerations limit the flume installation to two, 4 m wide structures placed next to each other. If the river is sufficiently wide, culvert flumes, which can withstand the passage of excavation equipment, can be utilized to provide additional flow bypass capacity.

- Figure 28 illustrates flume and pump bypass systems in winter conditions. It is noted that the hydraulic capacity of the flume is a function of its size, but equally important, of the details of its installation method – in the example shown, the large diversion dams partially blocked the inlet of the flume. As winter flows are subject to little change, pumps are well suited for flows less than 1.0 m^3/s although they require continuous operation during cold temperatures.

- Figure 29 presents schematics and typical detailed construction specifications for flow isolation methods. One of the keys to a successfully completed crossing is to ensure an adequate sediment settling pond or natural area is constructed or located. Techniques that can be utilized to reduce the size of the settling pond required are:
  - clean water seepage through or under the diversion dams should be isolated from the ditch area and returned directly to the river; and
  - minimizing the dewatering of the ditch area as the rate of seepage into the ditch line is directly proportional to the drawdown of the water level in the ditch area.

- Figure 30 illustrates typical bank restoration measures. If bank armoring is required to ensure the integrity of the crossing, a designed armor using select riprap or gabion baskets, or other armor options, will be required. In many cases utilizing naturally available boulders from the excavated bank and adjacent right-of–way area will provide a well restored and armored bank that can withstand most flow conditions.

- Figure 31 illustrates bed restoration techniques for bedrock and non-bedrock conditions. In the majority of crossings, replacing the excavated material is adequate from a technical and environmental viewpoint. For high velocity bedrock crossings, select backfill may be required to ensure

1. Dry, frozen trench with nearly vertical walls. Little or no flow into the ditch walls. Even minor ice buildup is a design concern for a hot oil pipeline (re: thawing and settlement) but less of a concern for a chilled/ambient temperature gasline.

2. Frozen trench with nearly vertical walls. Ice cover and water in the ditch during construction at -40°.

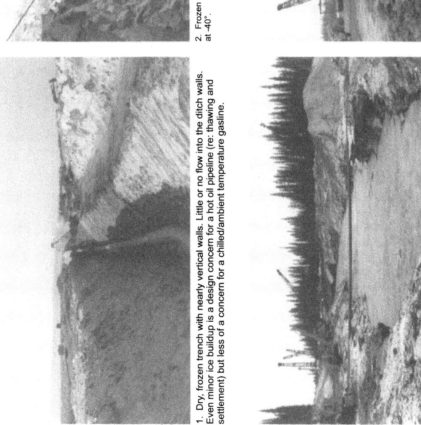

3. Frozen surface underlain by thawed material due to near surface spring-fed flow. Condition results in ditch instabilities and a significant ditch width which requires pipe placement via "pull-in" or by using backhoes rather than sidebooms.

4. Large wet ditch. Low flow. Achieving and maintaining ditch depth a challenge.

Figure 26 Examples, Open Cut Construction in the Winter. (Alaska)

120

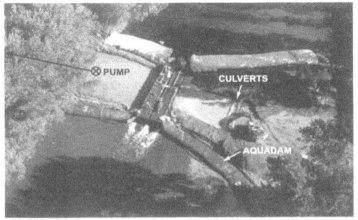

1. Twin flumes and culverts used to isolate the crossing work. Total flow bypassed = 10 m³/s.

2. Sediment settling ponds for the flow isolation in Photo 1. Wherever possible use natural depressions for the sediment pond. In the winter, snow and ice berms can be used to readily form sediment ponds.

3. Single flume, culverts and pumps. Total flow bypassed = 5 m³/s.

Figure 27 Examples, Flow Isolation Techniques, Summer (Canada)

1. Winter Flow Isolation using a flume and pumps. Upstream end. Partial blockage of flume inlet by the Aquadam necessitated the pumps to provide the required capacity and to reduce the head and seepage into the work area. (Also see Photo 2).

2. Winter Flow Isolation using a flume and pumps. Downstream end. Total flow bypassed = 3 m³/s.

3. Winter pumping using a sump upstream of the pipeline trench. Cold temperatures (to -40°C) required 24 hour operation of the pumps and extra pumps to minimize ice build up in the lines. Total flow bypassed with six pumps (6" - 8" in size) = 0.5 m³/s.

Figure 28 Examples, Flow Isolation Techniques. Winter (Canada)

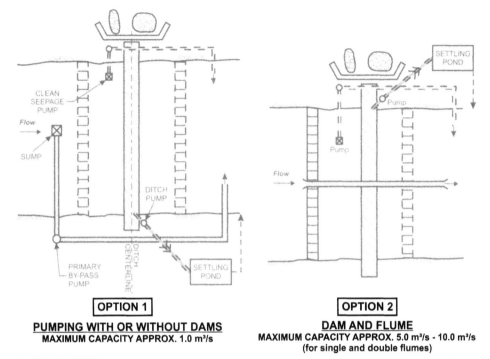

OPTION 1	OPTION 2
PUMPING WITH OR WITHOUT DAMS	**DAM AND FLUME**
MAXIMUM CAPACITY APPROX. 1.0 m³/s	MAXIMUM CAPACITY APPROX. 5.0 m³/s - 10.0 m³/s
	(for single and double flumes)

1.0 SCOPE OF DRAWING

    1.1 Scope - The schematics and notes are only to describe the specific and unique components and requirements of the Flow Isolation Technique. Other typical requirements for river crossings such as pump screens, re-fueling of pumps, in-stream and bank restoration etc. are covered on general drawings or in specifications.

2.0 FLOW DIVERSION

    2.1 Magnitude of flow in stream - Use historic or computed monthly flows.

    2.2 Diverting Natural versus Ditch Water - Flow unaffected by the work, may be diverted around the site and returned directly to the stream. Water pumped from the ditch shall be treated. Therefore maximizing the natural flow bypassed will minimize or may even eliminate the volume of flow from the ditch to be treated.

    2.3 Main Clean Water Pump capacity - Main pump to have a capacity equal to the stream's flow + 50%.

    2.4 Sump in the River - Use natural deep hole, if locally present, or excavate a sump.

    2.5 Outlet into the River - Layout and design to minimize erosion of riverbed. Use diffusers, riprap, geofabric materials or other form(s) of energy dissipation if necessary.

    2.6 Secondary Clean Water Pump - necessary if an upstream dam is used **and** significant seepage occurs through the dam or through the riverbed material.

    2.7 Pump in the Ditch
    - Needed if seepage water overflows the ditch and flows downstream.
    - Use this pump to drawdown the watertable in the ditch below the natural riverbed grade downstream of the ditch. Dewatering of the ditch is not required and in fact not desirable as it increases the seepage rate into the ditch.
    - The discharge of pumped seepage water from the ditch shall be filtered prior to release back into the river, via natural vegetation, or via natural or man-made sediment ponds, or via straw bales or silt fences. The optimum methodology required and selection depends on site conditions and rate of flow pumped.

3.0 DIVERSION DAM

    3.1 Purpose - If a sump or natural deep hole is adequate to "capture" the flow, a dam is not required - a preferred approach in most cases. A downstream dam is required only to prevent backup of the flow pumped around the site.

    3.2 Type
    - Place on existing streambed. Some leveling or removal of large boulders may be required to minimize seepage. Make the dam as watertight as practical.
    - Design options are:
      • concrete weights, blocks or barriers with an impermeable face of fabric or plywood across gaps or, sheet piling or,
      • water-filled rubber tubes,
      • sandbags (large or small) constructed of adequate strength fabric to enable handling (by hand or machines) without breakage.

    3.3 Placement/Removal - Place to minimize seepage and remove in a manner to minimize streambed disturbance. Boulders removed during construction of the dam to be replaced after removal of the dam.

Figure 29 Schematics of Flow Isolation Techniques and Water and Sediment Management Specifications.

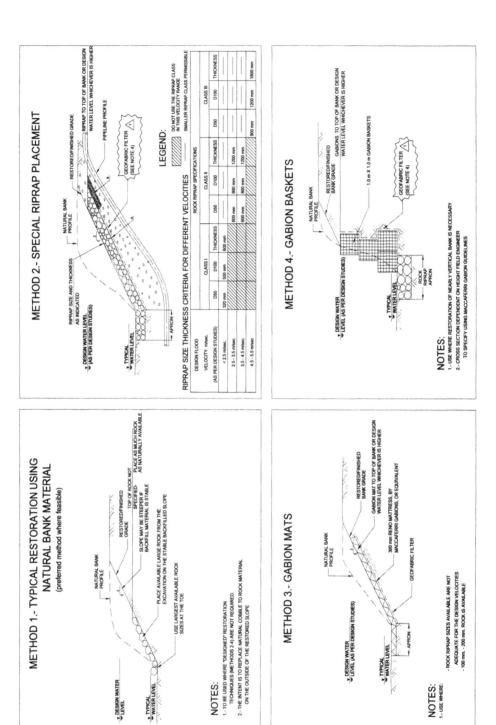

Figure 30 Typical Bank Restoration and Armoring Techniques.

124

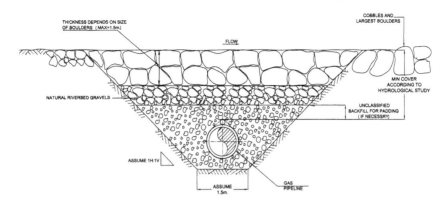

COBBLES OR BOULDER EXCAVATION
METHOD 1

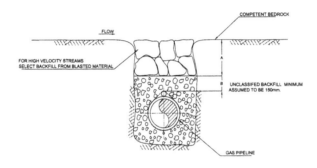

ROCK EXCAVATION
METHOD 2

GENERAL NOTES:

1.0 METHOD 0:

1.1 SORTING OF THE EXCAVATED OR BACKFILL MATERIAL NOT REQUIRED.

2.0 METHOD 1:

2.1 SORTING OF EXCAVATED MATERIAL REQUIRED, IF CROSSINGS HAVE A SURFICIAL LAYER OF LARGE MATERIAL, TO ENSURE THAT THE RESTORED STREAMBED HAS SCOUR AND HABITAT CHARACTERISTICS SIMILAR TO THAT OF THE NATURAL RIVER BED.

3.0 METHOD 2:

3.1 IN HIGH VELOCITY ROCK CHANNELS, THE SURFICIAL BACKFILL MUST BE OF AN ADEQUATE SIZE TO PREVENT IT'S EROSION AND PIPE EXPOSURE. THIS CAN GENERALLY BE READILY ACHIEVED BY SORTED BLASTED ROCK.

Figure 31 Typical Bed Restoration Techniques.

that, from an erodibility and thus pipe protection viewpoint, the backfill is adequate.

- Figure 32 illustrates examples of recently completed crossings in the Arctic. In floodplain areas, the usage of overfill is common considering the settlement that will occur with frozen backfill and a wet ditch. In the main channel area, overfill may not be feasible as it would affect winter low flow conditions. Fill settlement in the ditch line in the main channel may be pronounced in the first spring breakup; however, infilling will quickly occur as soon as the stream is thawed.

3.4 Operations

3.4.1 Monitoring Objectives

The design of the river crossings will be for the design flood and for the economic life of the pipeline and be based on the best available data and up-to-date methodologies. The construction will be in accordance with the design (Section 3.2), and construction techniques (Section 3.3).

River changes, either natural or construction-induced, may occur which were not predicted in the design phase. Or a flood greater-than-design may occur resulting in bank erosion or channel changes or the development and enlargement of new and existing subchannels respectively that were not predicted. Or the life of the pipeline may far exceed its anticipated economic life, thus requiring bank protection measures, particularly where the line is parallel to the river.

Therefore a thorough and consistent river monitoring program of the major crossings is an essential component of the sound operation of the pipeline.

3.4.2 Monitoring Requirements

The recommended monitoring program for major river crossings is as follows:

- Routine monitoring by operational personnel during the spring breakup or flood season. To facilitate this monitoring and particularly to readily observe changes in bank erosion from the air, the installation of large coloured survey stakes, is recommended.
- Re-profiling the floodplain and main channel one-year after construction is completed. For buried crossings that are not drilled (open cut or flow isolated), ditch settlement will occur during the first open water season. Until bed material is deposited into the settled ditch area – overfilling the ditch may not be desirable for construction (i.e. causing flooding of the work area) or environmental (i.e. causing a fish blockage point) reasons. In the floodplain areas where overfilling is probably permitted, settlement may be variable.

126

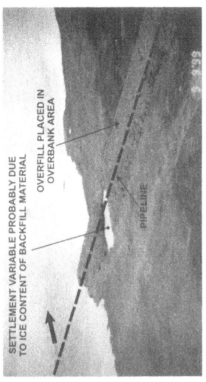

SETTLEMENT VARIABLE PROBABLY DUE TO ICE CONTENT OF BACKFILL MATERIAL

OVERFILL PLACED IN OVERBANK AREA

PIPELINE

1. Evidence of settlement of the ditchline for the entire width of the crossing. This is generally infilled during the first high runoff event.

2. Floodplain overfill.

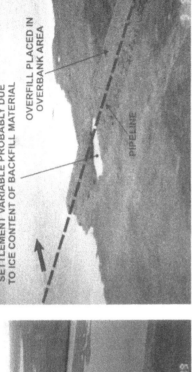

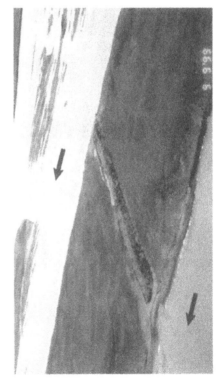

3. Floodplain overfill in a floodplain section which is overtopped infrequently. Depending on the degree of settlement, a function of the ice content of the fill, overfill may be present for a long period. Overfill is preferable to inadequate backfill which, in the example shown, could lead to flow between the two channels over and parallel to the pipeline.

4. Close-up of overfill in an active and lower floodplain area. The first significant flow will typically remove this overfill.

Figure 32 Examples, Overfill and Post Construction Settlement. (Arctic)

127

- Re-profile the crossing at specified intervals such as every five years (or as per regulatory requirements) plus after flood events which exceed a threshold value such as the 1:10 year event.
- Detailed assessments by river engineering specialists if a major flow has occurred and if visual or surveys indicate significant changes in the floodplain or main channel.

3.4.3 Monitoring Documentation

Two examples of monitoring formats, which illustrate an airphoto and profiles, outline issues and specific monitoring requirements and provide a summary of the results, are presented on Figures 33 and 34. The monitoring sheets should be updated annually for ease of reference by the operator and as necessary or desirable, for submission to the regulator.

3.4.4 Lessons Learned from Arctic River Crossings

Veldman (2002) summarizes the lessons learned from four major floods experienced during 25 years of operations of the 48-inch Trans Alaska oil pipeline. Some of these lessons are:

- The vast majority of bank erosion occurs during high flow late summer events. Comparative airphotos which "bracket" a high flow are thus useful for assessing past and potential river changes. On the other hand, numerous comparative airphotos that do not encompass a high flow event are not a good predictor of what could happen in the future.
- Bank erosion potential is greatest in the late summer when maximum thawing has occurred. Little bank erosion or development or enlargement of floodplain channels occurs during spring breakup when the ground is still frozen.
- As in the case of non-arctic rivers, local scour at bends, obstructions and rock banks is significant compared to general scour. In a single well-defined channel, flow pattern variability with time is minor thus the location and depth of historic scour holes is a predicative indicator of what could potentially affect the crossing. On a braided river the potential for flow pattern variability is high. Consequently, significant local scour can develop at a crossing where past scour was minor.
- In coarse gravel bed rivers, post-flood scour measurements are a reasonable indicator of maximum scour that occurred during the peak of the flood (scour measurements during the peak of a flood are generally not feasible for logistical and safety reasons). In fine-grained rivers, infilling of scour holes occurs rapidly after the passage of the flood peak thus post-flood surveys may not be a good indicator of the maximum scour that occurred during the flood.

MONITORING REQUIREMENTS

1. **Helicopter reconnaissance** especially during spring breakup.

   Important to assess what is happening upstream and downstream along the river and not solely at the pipeline crossing. Ice jams forming upstream ? or near the crossing ?

2. **Re-survey riverbed profile** at least every 5 years and following flows in excess of the 1:10 year event.

   In order to document trends/variability in the riverbed profile. If the crossing is open cut, recommend a re-survey in the first year after completion of the crossing

3. **Establish geodetic benchmarks** on both banks.

   To expedite future comparative surveys

4. **Install 3 monitoring posts** on each bank which are identifiable from a helicopter reconnaissance

   To facilitate a quick assessment of bank changes

SITE SPECIFIC RIVER ISSUES

1. **Bed Scour/Pipeline Cover** - due to high flows and/or ice breakup conditions.

2. **Bank Erosion/Set back to the Pipeline** - Due to major ice shoves during breakup. If the crossing is open cut, due to construction disturbance

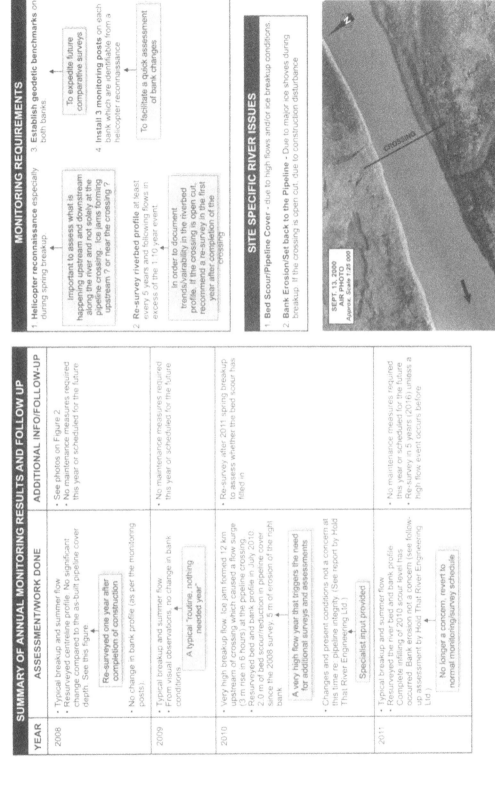

SEPT. 13, 2000
AIR PHOTO
Approx. Scale 1:25 000

CROSSING

SUMMARY OF ANNUAL MONITORING RESULTS AND FOLLOW UP

YEAR	ASSESSMENT/WORK DONE	ADDITIONAL INFO/FOLLOW-UP
2008	• Typical breakup and summer flow • Resurveyed centreline profile. No significant change compared to the as-built pipeline cover depth. See this figure. Re-surveyed one year after completion of construction • No change in bank profile (as per the monitoring posts).	• See photos on Figure 2 • No maintenance measures required this year or scheduled for the future
2009	• Typical breakup and summer flow • From visual observations, no change in bank conditions A typical "routine, nothing needed year"	• No maintenance measures required this year or scheduled for the future
2010	• Very high breakup flow. Ice jam formed 12 km upstream of crossing which caused a flow surge (3 m rise in 6 hours) at the pipeline crossing. • Resurveyed bed and bank profile in July 2010. 2.0 m of bed scour/reduction in pipeline cover since the 2008 survey. 5 m of erosion of the right bank A very high flow year that triggers the need for additional surveys and assessments • Changes and present conditions not a concern at this time re: pipeline integrity (See report by Hold That River Engineering Ltd.) Specialist input provided	• Re-survey after 2011 spring breakup to assess whether the bed scour has filled in
2011	• Typical breakup and summer flow • Resurveyed the river bed and bank profile. Complete infilling of 2010 scour level has occurred. Bank erosion not a concern (see follow-up assessment by Hold That River Engineering Ltd.) No longer a concern, revert to normal monitoring/survey schedule	• No maintenance measures required this year or scheduled for the future • Re-survey in 5 years (2016) unless a high flow event occurs before

Figure 33 Summary Monitoring Sheet, Major RIver Crossing.

SUMMARY OF ANNUAL MONITORING

YEAR	ASSESSMENT/WORK DONE	ADDITIONAL INFO/FOLLOW-UP
2008	• Re-serveyed centreline profile. See this Figure. • Typical breakup and summer flow. • Little or no visible change in the south floodplain.	• No maintenance measures required this year or scheduled for the future.
2009	• Extremely high, greater-than-design flow on August 13 which caused a significant increase in flow in the south floodplain. (see photos). • Re-surveyed centreline profile across the floodplain and main channel, see results this Figure (See report by specialist consultant, Hold That River Engineeering Ltd., October 12, 2009 for a river engineering assessment). • Riprap sill recommended to control erosion, and thus maintain cover depth.	• Sill constructed November 2009, see Drawing RC - 980 - 001, October 31, 2009, Rev. 0. This summary sheet therefore also serves as a tracking mechanism and summary document for all work associated with this crossing

Specialist input provided

MONITORING REQUIREMENTS

1. Helicopter reconnaissance especially noting changes in flow patterns upstream which include flow into the south floodplain.
2. Re-survey riverbed profile at least every 5 years and following flows in excess of the 1:10 year event.
3. Establish geodetic benchmarks on both banks.
4. Install 3 monitoring posts on each bank.

SITE SPECIFIC RIVER ISSUES

1. Bed Scour/Pipeline Cover - due to high flows or debris obstructions.
2. Bank Erosion - Due to high flows, or channel changes such as gravel bar deposition or formation of new channels, or flow deflection caused by debris.
3. Channel Changes - high spring ice levels or high summer floods which divert flow into the south floodplain causing the enlargement of the existing sub-channel.

Site specific issue for the Ochre River

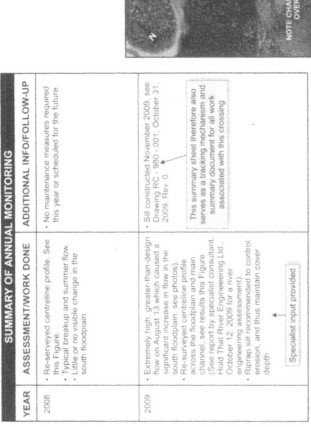

Figure 34 Summary Monitoring Sheet, River Crossing with a Floodplain.

- Ice shove up the riverbanks can, depending on ice, river and bank conditions, reach elevations significantly higher than even extreme design flood or mid-river ice levels. Bank armoring and restoration measures will need to be designed considering ice shove. Appurtenances such as valves (if projecting above ground) and cathodic protection stations at river crossings should be located outside the shove area or protected.
- Aufeis (icing) elevations are generally highly variable from year-to-year and from location to location. The generally accepted climatic reasons for producing maximum aufeis development – cold in late fall and low snowfall – are valid, however at certain crossings or floodplain/instream areas, the opposite conditions may generate maximum aufeis.
- In streams with significant aufeis development, the maximum water levels are generally experienced for normal breakup flow over the ice.

3.5 References

(1) Blench, T. "Mobile-Bed Fluviology." Edmonton, Alberta: The University of Alberta Press, 1969.

(2) Hydrocon Engineering (Continental) Ltd. "A Review of Scour Multiplication Factors for gravel-Bed Rivers." Calgary, Alberta, October 1983.

(3) Nanson and Hickin. "Channel Migration and Incision on the Beatton River: Journal of Hydraulic Engineering." 1983.

(4) Neill, C.R. "Guide to Bridge Hydraulics." Published for Roads and Transportation Association of Canda by University of Toronto Press, 1973.

(5) Veldman W.M. "Arctic Pipeline River Crossings, Design Trends and Lessons Learned." Prepared for the American Society of Civil Engineers, Specialty Conference on Pipelines in Adverse Environments, San Diego, California, November 1983.

(6) Veldman W.M. "Lessons Learned for river Crossing Designs from Four Major Floods Experienced along the Trans Alaska Pipeline." ASCE, Cold Regions Engineering, Anchorage, Alaska, May 2002.

4 Horizontal Directional Drilling

4.1 Introduction

The horizontal directional drilling (HDD) process represents a significant improvement over traditional cut and cover methods for installing pipelines beneath obstacles that warrant specialized construction attention. In order for these advantages to be realized, creative engineering must be applied in advance of construction. Design engineers should have a working knowledge of the process in order to produce designs that can be efficiently executed in the field. This chapter describes the fundamentals involved in designing a drilled installation as well as contractual and construction monitoring considerations. Topics covered include site investigation, drilled path design, temporary workspace requirements, drilling fluids, pipe specification, contractual considerations, and construction monitoring.

4.2 The Horizontal Directional Drilling Process

Installation of a pipeline by HDD is generally accomplished in three stages as shown in Figure 4.1. First, a small diameter pilot hole is drilled along a designed directional path. Next, this pilot hole is enlarged to a diameter that will accommodate the pipeline. Finally, the pipeline is pulled into the enlarged hole.

4.2.1 Pilot Hole

Pilot hole directional control is achieved by using a non-rotating drill string with an asymmetrical leading edge. The asymmetry of the leading edge creates a steering bias while the non-rotating aspect of the drill string allows the steering bias to be held in a specific position while drilling. If a change in direction is required, the drill string is rolled so that the direction of bias is the same as the desired change in direction. Leading edge asymmetry is typically accomplished with a bent sub or bent motor housing located several feet behind the bit as illustrated in Figure 4.2.

In soft soils, drilling progress is typically achieved by hydraulic cutting with a jet nozzle. If hard spots are encountered, the drill string may be rotated to drill without directional control until the hard spot has been penetrated. Mechanical cutting action required for harder soils is provided by a positive displacement mud motor which converts hydraulic energy from drilling fluid to mechanical energy at the drill bit. This allows for bit rotation without drill string rotation.

The actual path of the pilot hole is monitored during drilling using a steering tool positioned near the bit. The steering tool provides continuous readings of the inclination and azimuth at the leading edge of the drill string. These readings, in conjunction with measurements of the distance drilled, are used to calculate the horizontal and vertical coordinates of the steering tool relative to the initial entry point on the surface. When the drill bit penetrates the surface at the exit point opposite the horizontal drilling rig, the pilot hole is complete.

PILOT HOLE

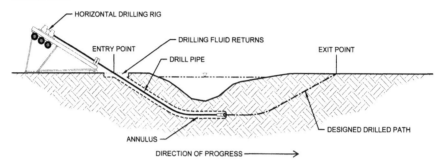

PREREAMING

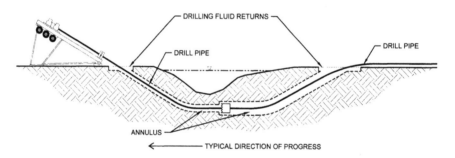

PULLBACK

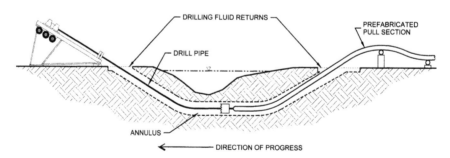

Figure 4.1 – The HDD Process

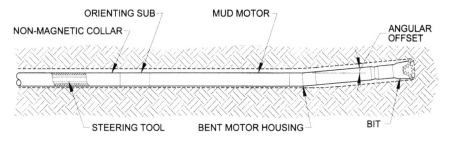

Figure 4.2 – Bottom Hole Assembly

Figure 4.3 – Mud Motor and Bit

4.2.2 Prereaming

Enlargement of the pilot hole is typically accomplished using prereaming passes prior to pipe installation. Reaming tools generally consist of a circular array of cutters and drilling fluid jets and are often custom made by contractors for a particular hole size or type of soil. Examples of different types of reaming tools are shown in Figures 4.4, 4.5, and 4.6.

Figure 4.4 – Flycutter

Figure 4.5 – Barrel Reamer

Figure 4.6 – Hole Opener

For a typical prereaming pass, a reamer attached to the drill string at the exit point is rotated and drawn to the drilling rig thus enlarging the pilot hole. Drill pipe is added behind the reamer as it progresses toward the drill rig to ensure that a string of pipe is always maintained in the drilled hole. It is also possible to ream away from the drill rig, in which case a reamer fitted into the drill string at the rig is rotated and advanced away from it.

4.2.3 Pullback

Pipe installation is accomplished by attaching the prefabricated pipeline pull section behind a reaming assembly at the exit point, then pulling the reaming assembly and pull section back to the drilling rig. This is undertaken after completion of prereaming or, for smaller diameter lines in soft soils, directly after completion of the pilot hole. A swivel is utilized to connect the pull section to the reaming assembly to minimize torsion transmitted to the pipe. The pull section is supported using some combination of roller stands, pipe handling equipment, or a flotation ditch to minimize tension and prevent damage to the pipe.

137

Figure 4.7 – Pipe Handling Equipment

Figure 4.8 – Pull Section Breakover

4.3 Site Investigation

The first step in accomplishing an HDD installation is to investigate the site at which the work will be undertaken. An appropriate site investigation will include both surface and subsurface surveys. Although each survey may be performed by a different specialized engineering consultant, it is important that the results be integrated onto a single plan and profile drawing which will be used to price, plan, and execute the crossing. Since this drawing will also be used to produce the working profile that will be the contractor's basis for downhole navigation, accurate measurements are essential.

4.3.1 Surface Survey

A topographic survey should be conducted to accurately describe the working areas where construction activities will take place. Both horizontal and vertical control must be established for use in referencing hydrographic and geotechnical data. A typical survey should include overbank profiles on the crossing centerline extending from approximately 150 feet (46 meters) landward of the entry point to the length of the prefabricated pull section landward of the exit point. Survey ties should also be made to topographic features in the vicinity of the crossing.

For significant waterways, a hydrographic survey will be required to accurately describe the bottom contours. Typically, it should consist of fathometer readings along the crossing centerline and approximately 200 feet (61 meters) upstream and downstream. This scope can be expanded to include more upstream and/or downstream ranges if additional data is required to analyze future river activity.

4.3.2 Subsurface Survey

A subsurface survey for an HDD installation should define the geological characteristics and engineering properties of the subsurface material through which the drilled path will pass. It should include both a review of existing geological information and the assembly of site-specific data obtained from exploratory borings. The extent of the subsurface survey should be governed by practical economic limits.

Existing geological data should be reviewed to determine the general subsurface conditions at the specified location. A site-specific geotechnical investigation should also be conducted to confirm the probable subsurface conditions through which the crossing will be installed. The number and location of borings, as well as the use of other exploratory techniques, should be based on site-specific conditions taking into account the preliminary drilled path design. Borings should be located approximately 50 feet (15 meters) off of the crossing centerline and should extend approximately 30 feet (9 meters) below the deepest crossing penetration depth.

The sampling interval and technique should be based on site-specific conditions and designed to accurately describe the subsurface material. If rock is encountered, the borings should at a minimum penetrate the rock to a depth sufficient to confirm that it is bedrock, and preferably should extend beyond the HDD crossing's penetration depth to provide detailed information about the bedrock properties. The following data should be obtained from the borings:

- Standard classification of soils
- Gradation curves for granular soils
- Standard penetration test (SPT) values where applicable
- Cored samples of rock with rock quality designation (RQD) and percent recovery
- Unconfined compressive strength for rock samples
- Mohs Hardness for rock samples

The results of the geotechnical investigation should be presented in the form of a geotechnical report containing a brief description of the local geology, pertinent engineering analysis, boring logs, laboratory test results, and a profile of the anticipated subsurface conditions along the proposed HDD alignment.

4.4 Drilled Path Design

To maximize the advantages offered by HDD, primary design consideration should be given to defining the obstacle to be crossed. For example, a river is a dynamic entity. Not only should its width and depth be considered, the potential for bank migration and scour during the design life of the crossing should also be taken into account. It should be remembered that flexibility in locating a pipeline to be installed by HDD exists not only in the horizontal plane but in the vertical plane as well.

For the majority of drilled installations, there are six parameters that define the location and configuration of the drilled path. These are the entry and exit points, the entry and exit angles, the P.I. elevation, and the radius of curvature. These parameters, or their limiting values, should be specified on the contract plan & profile drawing. The relationship of these parameters to each other is illustrated in Figure 4.9.

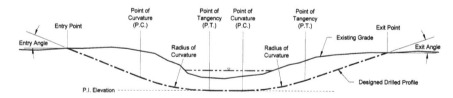

Figure 4.9 – HDD Design Terminology

4.4.1 Entry and Exit Points

The entry and exit points are the endpoints of the designed drilled segment on the ground surface. The drilling rig is positioned at the entry point and the pipeline is pulled into the drilled hole from the exit point. The relative locations of the entry and exit point, and consequently the direction of pilot hole drilling, reaming, and pullback, should be established by the site's geotechnical and topographical conditions. When choosing the relative locations of the entry and exit point, it is important to note that steering precision and drilling effectiveness are greater near the drilling rig. Where possible, the entry point should be located close to anticipated adverse subsurface conditions. An additional consideration when selecting entry and exit point locations is the availability of workspace for pull section fabrication.

4.4.2 Entry and Exit Angles

Entry angles should be held between 8 and 20 degrees with the horizontal. These boundaries are primarily due to equipment limitations as HDD rigs are often manufactured to operate between 10 and 18 degrees. Exit angles should be designed to facilitate breakover support during pullback. That is, the exit angle should not be so steep that the pull section must be severely elevated in order to guide it into the drilled hole. This will generally be less than 10 degrees for larger diameter lines.

4.4.3 P.I. Elevation

The P.I. elevation defines the depth of cover that the installation will provide. Adequate cover should be provided to maintain crossing integrity over its design life. Typically, HDD crossings should be designed to provide around 25 feet (7.6 meters) of cover and only in rare cases should less than 15 feet (4.6 meters) of cover be provided. This aids in reducing inadvertent drilling fluid returns and provides a margin for error in existing grade elevation and pilot hole calculations. Designed depth of cover is typically increased for installations beneath sensitive obstacles such as major waterways, highways, and railroads. Geotechnical factors should also be considered when selecting the vertical position of the pipeline.

4.4.4 Radius of Curvature

The industry standard design radius of curvature for circular bends used in HDD installations is determined by the following formula:

$$R = 1200(D_{nom})$$

Where:　　　　R　=　Radius of curvature of circular bends

　　　　　　　D_{nom}　=　Nominal diameter of the pipe

141

This relationship has been developed over a period of years in the HDD industry and is based on experience with constructability as opposed to any theoretical analysis. A lesser, minimum allowable radius should be specified in the pilot hole tolerances to provide the contractor some flexibility during construction. This minimum allowable radius should be determined through an analysis of installation and operating stresses.

4.5 Temporary Workspace Requirements

4.5.1 Horizontal Drilling Rig

A typical large horizontal drilling spread can be moved onto a site in approximately seven tractor-trailer loads. A workspace of 150 feet (46 meters) by 250 feet (76 meters) is adequate for most operations. If necessary, an HDD spread may be assembled in a minimal workspace of 60 feet (18 meters) by 150 feet (46 meters), however this minimal workspace will restrict the size and capacity of the drilling rig. Space requirements will vary depending on the make and model of the rig and how the various components are positioned. However, the locations of the principal components of the spread (rig ramp, drill pipe, and control trailer) are fixed by the entry point. The rig ramp must be positioned in line with the drilled segment and typically less than 25 feet (7.6 meters) back from the entry point. The control trailer and drill pipe must be positioned adjacent to the rig.

A typical horizontal drilling rig site plan is shown in Figure 4.10. A photograph of a large rig HDD spread is included as Figure 4.11.

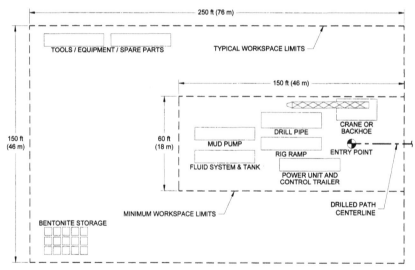

Figure 4.10 – Typical Horizontal Drilling Rig Site Plan

142

Figure 4.11 – Large Horizontal Drilling Rig Spread

Workspace for the horizontal drilling rig must be cleared and graded level. Equipment is typically supported on the ground surface, although timber mats may be used where soft ground is encountered. For marine locations, it is possible to operate off of a barge.

4.5.2 Pull Section Fabrication

Pull section fabrication is accomplished using the same construction methods that are used to lay a pipeline; therefore, similar workspace is required. The drilled segment exit point controls the location of pull section fabrication workspace. Space must be available to allow the pipe to be fed into the drilled hole. It is preferable to have workspace aligned with the drilled segment, extending back from the exit point the length of the pull section plus 200 feet (61 meters). This will allow the pull section to be prefabricated in one continuous length prior to installation. If space is not available, the pull section may be fabricated in two or more segments which are welded or fused together during installation. However, delays associated with joining multiple pipe segments during pullback increase the risk of the pipe becoming stuck in the hole.

Workspace for pull section fabrication should generally be around 50 feet wide, similar to what is required for conventional pipeline construction. Additional temporary workspace should be provided in the immediate vicinity of the exit point to facilitate personnel and equipment supporting drilling operations. Pull

143

section workspace must be cleared but need not be graded level. Equipment is typically supported on the ground surface. Timber mats may be used where soft ground is encountered.

A typical pull section fabrication site plan is shown in Figure 4.12. A photograph showing pull section staging operations is included as Figure 4.13.

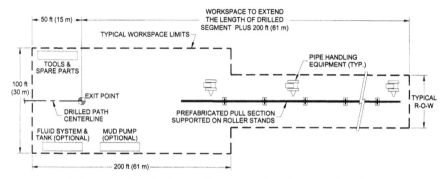

Figure 4.12 – Typical Pull Section Fabrication Site Plan

Figure 4.13 – Pull Section Staging Operations

144

4.6 Drilling Fluids

The primary impact of HDD on the environment revolves around the use of drilling fluids. Where regulatory problems are experienced, the majority of concerns and misunderstandings are associated with drilling fluids. An awareness of the function and composition of HDD drilling fluids is imperative in producing a permittable and constructible HDD design. A detailed discussion of drilling fluids relative to HDD installations can be found in *Drilling Fluids in Pipeline Installation by Horizontal Directional Drilling, A Practical Applications Manual.*

4.6.1 Drilling Fluid Flow Schematic

Drilling fluid is used in all phases of the HDD process. The schematic diagram below shows the relationship of the elements in a typical HDD drilling fluid system.

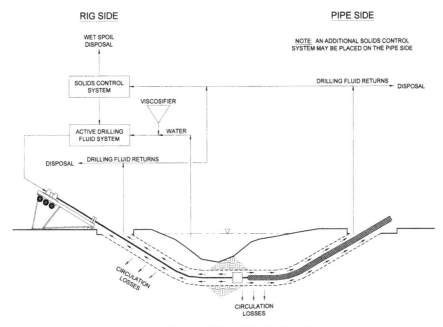

Figure 4.14 – HDD Drilling Fluid Flow Schematic

4.6.2 Functions of Drilling Fluid

The principal functions of drilling fluid in HDD pipeline installation are listed below:

- **Transportation of Spoil.** Drilled spoil, consisting of excavated soil or rock cuttings, is suspended in the fluid and carried to the surface by the fluid stream flowing in the annulus between the wall of the hole and the pipe.

- **Cooling and Cleaning of Cutters.** High velocity fluid streams directed at the cutters remove drilled spoil build-up on bit or reamer cutters. The fluid also cools the cutters.

- **Reduction of Friction.** Friction between the pipe and the wall of the hole is reduced by the lubricating properties of the drilling fluid.

- **Hole Stabilization.** The drilling fluid stabilizes the drilled or reamed hole. This is critical in HDD pipeline installation as holes are often in soft soil formations and are uncased. Stabilization is accomplished by the drilling fluid building up a wall cake and exerting a positive pressure on the hole wall. Ideally, the wall cake will seal pores and produce a bridging mechanism to hold soil particles in place.

- **Transmission of Hydraulic Power.** Power required to turn a bit and mechanically drill a hole is transmitted to a downhole motor by the drilling fluid.

- **Hydraulic Excavation.** Soil is excavated by erosion from high velocity fluid streams directed from jet nozzles on bits or reaming tools.

- **Soil Modification.** Mixing of the drilling fluid with the soil along the drilled path facilitates installation of a pipeline by reducing the shear strength of the soil to a near fluid condition. The resulting soil mixture can then be displaced as a pipeline is pulled into it.

4.6.3 Composition of HDD Drilling Fluid

The major component of drilling fluid used in HDD pipeline installation is fresh water obtained at the crossing location. In order for water to perform the functions listed above, it is generally necessary to modify its properties by adding a viscosifier. The viscosifier used almost exclusively in HDD drilling fluids is naturally occurring clay in the form of bentonite mixed with small amounts of extending polymers to increase its yield (high yield bentonite).

Increasing the yield of bentonite allows more drilling fluid to be produced with less viscosifier (dry bentonite). For example, Wyoming bentonite yields in excess of 85 barrels (1 barrel = 42 gallons) of drilling fluid per ton of dry viscosifier. Addition of polymers to produce high yield bentonite can increase the yield to more than 200 barrels of fluid per ton of viscosifier. Typical HDD drilling fluids are composed of less than 2% high yield bentonite by volume with the remaining components being water and drilled spoil. Solids control equipment should be utilized to remove drilled spoil from the fluid to the extent practical, maintaining total solids (high yield bentonite and drilled spoil) at around 6% by volume. This equates to a drilling fluid density of approximately 9 pounds per gallon.

4.6.4 Material Descriptions

Generic descriptions of available HDD drilling fluid components follow. Manufacturers and service companies can be contacted for details on specific products.

4.6.4.1 Bentonite

Bentonite has a high affinity for water and swells to as much as 20 times its dry state when immersed in fresh water. Hydrated bentonite has excellent suspending characteristics and reduces filtrate loss. The term bentonite includes any member of the general montmorillonite group, which includes montmorillonite, beidellite, nontronite, hectorite, and saponite. The bentonite normally used in drilling fluids comes from Wyoming and South Dakota and is principally sodium montmorillonite (7) (12).

4.6.4.2 Attapulgite

Attapulgite is primarily used to make drilling fluid with saltwater where freshwater is not available. Viscosity produced by attapulgite is purely mechanical; hydration forces are not involved. Chemically, attapulgite is a hydrous magnesium silicate. Viscosity is produced by the disintegration of its crystalline structure into numerous needle-like particles. These particles tend to stack up, creating a brush-heap effect and providing viscosity (12).

4.6.4.3 Polymers

Polymers used in drilling fluids include both natural and synthetic compounds.

Xanthan gum is a high molecular-weight polysaccharide used to produce viscosity in freshwater and saltwater muds. It is manufactured using a controlled fermentation process and provides viscosity, yield value, and gel strength in saline waters without benefit of other colloidal materials such as bentonite. A preservative is required to prevent bacterial degradation (12). Xanthan gum is environmentally benign and is approved by the FDA for use in food products.

Polyanionic cellulose is used primarily as a fluid-loss reducer for freshwater and saltwater muds, but also acts as a viscosifier in these systems. It is environmentally benign and is not subject to bacterial degradation (12).

Sodium carboxymethyl cellulose (CMC) is primarily a fluid-loss reducer but also produces viscosity in freshwater and saline muds whose salt content does not exceed 50,000 mg/l. CMC is a long-chain molecule that can be polymerized to produce different molecular weights and, in effect, different viscosity grades. CMC is generally available in a high or low viscosity type. Either grade provides effective fluid-loss control. It is not subject to bacterial degradation (12). CMC is environmentally benign and is considered biodegradable. Cellulose derivatives, such as CMC, are used in foods.

147

4.6.5 Inadvertent Returns

HDD involves the subsurface discharge of drilling fluids. Once discharged downhole, drilling fluid is uncontrolled and will flow in the path of least resistance. This can result in dispersal into the surrounding soils or discharge to the surface at some random location, which may not be a critical problem in an undeveloped location. However, in an urban environment or a high profile recreational area, inadvertent returns can be a major problem. In addition to the obvious public nuisance, drilling fluid flow can buckle streets or wash out embankments.

Drilling parameters may be adjusted to maximize drilling fluid circulation and minimize the risk of inadvertent returns. However, the possibility of lost circulation and inadvertent returns cannot be eliminated. Contingency plans addressing possible remedial action should be made in advance of construction and regulatory bodies should be informed.

Inadvertent returns are more likely to occur in less permeable soils with existing flow paths. Examples are slickensided clay or fractured rock structures. Coarse grained, permeable soils exhibit a tendency to absorb circulation losses. Manmade features, such as exploratory boreholes or piles, may also serve as conduits to the surface for drilling fluids.

Inadvertent returns in a waterway and a sinkhole resulting from inadvertent returns are shown in the following two photographs.

Figure 4.15 – Inadvertent Returns in a Waterway

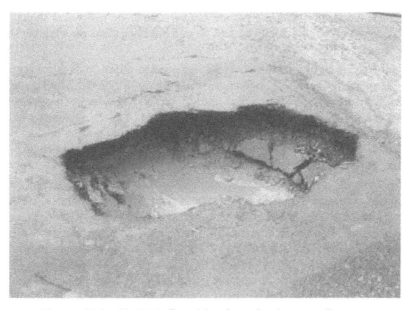

Figure 4.16 – Sinkhole Resulting from Inadvertent Returns

Research projects have been conducted in an attempt to identify the mechanisms that cause inadvertent returns and develop analytical methods for use in predicting their occurrence. Efforts have centered on predicting hydrofracturing. These programs have met with limited success in providing a reliable prediction method. Engineering judgment and experience must be applied in utilizing the hydrofracturing model to predict the occurrence, or nonoccurrence, of inadvertent returns.

4.6.6   Excess Drilling Fluid and Drilled Spoil Disposal

The HDD construction process generates drilled spoil and excess drilling fluid that must be disposed of properly. In addressing disposal regulations, it is important to remember that HDD drilling fluid is typically composed only of water, high yield bentonite, and drilled spoil. The major component of the fluid is water which is generally obtained either from the waterway being crossed or a municipal source. On most HDD crossings, the only foreign material introduced to the location is naturally occurring bentonite with a small amount of extending polymers. Applicable disposal regulations should be similar to those governing sedimentation and erosion control or general excess construction spoil disposal.

4.6.6.1   Recirculation

The first step in effectively dealing with excess drilling fluid disposal is to minimize the excess. This is accomplished by recirculating drilling fluid returns to

149

the extent practical. Collected surface returns should be processed through a solids control system that removes spoil from the drilling fluid allowing the fluid to be reused.

The basic solids control method used for water-based HDD fluids is mechanical separation. Solids control systems are not 100% efficient. That is, spoil discharged is not dry and totally free of drilling fluid and fluid discharged is not totally free of drilled spoil. The consistency of the spoil discharged from a solids control system may be similar to ready mix cement or a very viscous drilling fluid depending on factors such as the subsurface material being penetrated, drilling fluid properties, and the skill of the operator.

Recirculation on an HDD waterway crossing is complicated by the fact that a significant portion of the drilling fluid returns occur at the exit point on the bank opposite of the drilling rig. As a result, either two drilling fluid systems must be utilized or drilling fluid returns must be transported from the exit point to the drilling rig location. Transportation of drilling fluid returns can be accomplished by truck, barge, or a temporary recirculation line drilled beneath the bottom of the waterway. Which system is most advantageous will be determined by site-specific conditions. In some cases, temporary recirculation lines have been laid directly on the bottom of the waterway. However, this procedure involves the risk of rupture and resulting discharge of drilling fluid into the waterway.

4.6.6.2 Land Farming

Land farming provides an efficient and effective way to dispose of excess drilling fluid or drilled spoil. This disposal method involves distributing the excess drilling material evenly over an open area and mechanically incorporating it into the soil. The character and amount of the excess material will dictate the degree of mechanical tilling. Small quantities of whole fluid will dissipate with little or no tilling. If large quantities of fluid or wet spoil are involved, a significant tilling effort may be required to ensure that the drilling fluid components don't form a dry crust and remain in a semi-solid state over an extended period of time. The condition of the land farming site should typically be governed by standard construction clean-up and site restoration specifications.

4.6.6.3 Dewatering

Dewatering of excess drilling fluid offers significant environmental benefits. However, dewatering systems are expensive. Because of this, they are rarely, if ever, used on HDD installations. The objective of a dewatering system is to remove all of the solids from the drilling fluid. Solids removed include not only drilled spoil, but also the viscosifier that has been added to enhance fluid properties (typically high yield bentonite). Dewatering can take place during drilling operations or after completion. If dewatering is concurrent with drilling, processed water should be returned to the active drilling fluid system for mixing and reuse. If dewatering follows construction, processed water should be discharged in accordance with local regulations.

Solids produced by an appropriate dewatering system should be "dry". That is, they can be handled with standard earth moving and hauling equipment. They should be disposed of in accordance with local regulations. Typically, this will be in a similar manner to general excavation spoil. However, regardless of the disposal requirements for dry spoil, eliminating the need to dispose of liquids and minimizing the total mass of material requiring disposal can significantly reduce disposal costs.

4.7 Pipe Specification

Load and stress analysis for an HDD pipeline installation is different from similar analyses of conventionally buried pipelines because of the relatively high tension loads, bending, and external fluid pressures acting on the pipeline during the installation process. In some cases these loads may be higher than the design service loads. Pipe properties such as yield strength and wall thickness must be selected such that the pipeline can be both installed and operated within customary risks of failure.

In most cases, the wall thickness and specified minimum yield strength of the pipe will be determined by applicable codes and regulations. However, loads and stresses imposed during installation should be reviewed and analyzed in combination with operating stresses to insure that acceptable limits are not exceeded. Analysis of the loads and stresses that govern pipe specification can most easily be accomplished by breaking the problem into two distinct events: installation and operation.

4.7.1 Installation Loads

During HDD installation, a pipeline segment is subjected to tension, bending, and external pressure as it is pulled through a prereamed hole. The stresses in the pipe and its potential for failure are a result of the interaction of these loads (9) (11). In order to determine if a given pipe specification is adequate, HDD installation loads must first be estimated so that the stresses resulting from these loads can be calculated. A description of HDD installation loads and a method for estimating these loads follows.

4.7.1.1 Tension

Tension on the pull section results from three primary sources: frictional drag between the pipe and the wall of the hole, fluidic drag from viscous drilling fluid surrounding the pipe, and the effective (submerged) weight of the pipe as it is pulled through the hole. In addition to these forces that act within the drilled hole, frictional drag from the portion of the pull section remaining on the surface (typically supported on rollers) also contributes to the tensile load on the pipe.

Other loads that the horizontal drilling rig must overcome during pullback result from the length of the drill string in the hole and the reaming assembly that precedes the pull section. These loads don't act on the pull section and therefore have no impact on pipe stresses. Nonetheless, if a direct correlation with the overall rig force is desired, loads resulting from the reaming assembly and drill string must be estimated and added to the tensile force acting on the pull section.

4.7.1.1.1   Frictional Drag

Frictional drag between the pipe and soil is determined by multiplying the bearing force that the pull section exerts against the wall of the hole by an appropriate coefficient of friction. A reasonable value for coefficient of friction is 0.30 for a pipe pulled into a reamed hole filled with drilling fluid (6). However, it should be noted that this value can vary with soil conditions. A very wet mucky soil may have a coefficient of friction of 0.1 while a rough and dry soil (unlikely in an HDD installation) may have a coefficient of friction of 0.8.

For straight segments, the bearing force can be determined by multiplying the segment length by the effective unit weight of the pipe and resolving this force into a radial component based on the angle of the segment. For curved segments, calculation of the bearing force is more complicated since additional geometric variables must be considered along with the stiffness of the pipe.

4.7.1.1.2   Fluidic Drag

Fluidic drag resulting from drilling fluid surrounding the pipe is determined by multiplying the external surface area of the pipe by an appropriate fluid drag coefficient. A reasonable value for fluidic drag coefficient is 0.025 pounds per square inch (15). The external surface area of any segment defined in the drilled path model can easily be determined based on the segment's length and the outside diameter of the pull section.

4.7.1.1.3   Effective Weight of Pipe

The effective weight of the pipe is the unit weight of the pull section minus the unit weight of any drilling fluid displaced by the pull section. This is typically expressed in pounds per foot. The unit weight of the pull section includes not only the product pipe, but also its contents (ducts, internal water used for ballast, etc.) and external coatings if substantial enough to add significant weight (i.e. concrete coating). Calculating the weight of drilling fluid displaced by the pull section requires that the density of the drilling fluid either be known or assumed. For HDD installations, drilling fluid density will range from approximately 8.9 pounds per gallon to approximately 11 pounds per gallon (1) (10). Where use of a high end value for fluid density is warranted for a conservative analysis, 12 pounds per gallon represents a reasonable upper limit.

The pulling load from the effective weight of the pipe is determined by resolving the effective weight into an axial component based on the angle of the segment.

4.7.1.2 Bending

The pull section is subjected to elastic bending as it is forced to negotiate the curvature of the hole. For a pipe with welded or fused joints, this induces a flexural stress in the pipe that is dependent upon the drilled radius of curvature. For steel pipe, the relatively rigid material's resistance to bending also induces a normal bearing force against the wall of the hole. These normal forces influence the tensile load on the pipe as a component of frictional drag.

4.7.1.3 External Pressure

During HDD installation, the pull section is subjected to external pressure from four sources: 1) hydrostatic pressure from the weight of the drilling fluid surrounding the pipe in the drilled annulus, 2) hydrokinetic pressure required to produce drilling fluid flow from the reaming assembly through the reamed annulus to the surface, 3) hydrokinetic pressure produced by surge or plunger action involved with pulling the pipe into the reamed hole, and 4) bearing pressure of the pipe against the hole wall as the pipe is forced to conform to the drilled path.

Hydrostatic pressure is dependent upon the height of the drilling fluid column acting on the pipe and the density of the drilling fluid that surrounds the pipe. Drilling fluid density values are discussed in Section 4.7.1.1.3. The height of the drilling fluid column at any given location along the drilled path is typically equal to the elevation difference between that location and the point at which there is no drilling fluid in the reamed hole. Typically, but not always, drilling fluid extends to the entry or exit point, whichever is lower.

Hydrokinetic pressure required to produce drilling fluid flow can be calculated using annular flow pressure loss formulas. These results are dependent on detailed drilling fluid properties, flow rates and hole configuration and, because of uncertainties involving these parameters, often require a substantial application of engineering judgment to determine a reasonable value. In most cases, annular flow during pullback is low velocity with low pressure losses.

Hydrokinetic pressure due to surge or plunger action and hole wall bearing pressure cannot be readily calculated and must be estimated using engineering judgment and experience.

4.7.2 Installation Stresses

A thorough design process requires examination of the stresses that result from each individual installation loading condition as well as an examination of the combined stresses that result from the interaction of these loads.

153

4.7.2.1 Tensile Stress (f_t)

The tension imposed on a circular pipe during installation by HDD is assumed to act through the centroid of the cross section and therefore is uniformly distributed over the cross section. The tensile stress is determined by dividing the tension by the cross sectional area. The maximum allowable tensile stress imposed on a steel pull section during installation should be limited to 90% of the pipe's specified minimum yield strength (6).

4.7.2.2 Bending Stress (f_b)

Bending stress resulting from a rigid steel pipe being forced to conform to the drilled radius of curvature can be calculated using the following equation (18).

$$f_b = (ED)/(2R)$$

Where:

f_b = longitudinal stress resulting from bending in pounds per square inch
E = modulus of elasticity for steel, 29,000,000 pounds per square inch (17)
D = outside diameter of the pipe in inches
R = Radius of curvature of circular bends in inches

Bending stress imposed on a steel pull section during installation should be limited as follows (6). These limits are taken from design criteria established for tubular members in offshore structures and are applied to HDD installation because of the similarity of the loads on pipe (2).

$$F_b = 0.75F_y \qquad \text{for} \qquad D/t \le 1,500,000/F_y$$

$$F_b = (0.84 - (1.74F_y D)/(Et))F_y \qquad \text{for} \qquad 1,500,000/F_y < D/t \le 3,000,000/F_y$$

$$F_b = (0.72 - (0.58F_y D)/(Et))F_y \qquad \text{for} \qquad 3,000,000/F_y < D/t \le 300,000$$

Where:

F_b = maximum allowable bending stress in pounds per square inch
F_y = pipe specified minimum yield strength in pounds per square inch
t = pipe wall thickness in inches

In the HDD industry, it is standard practice to design circular sag bends for steel pipelines at a radius of curvature of 1,200 times the nominal diameter of the product pipe (refer to Section 4.4.4). As stated previously, this relationship is based on experience with constructability as opposed to pipe stress limitations. Typically, the minimum radius determined using the stress-limiting criterion presented above will be substantially less than 1,200 times the nominal diameter.

For this reason, bending stress limits rarely govern geometric drilled path design but are applied, along with other stress limiting criteria, in determining the minimum allowable radius of curvature.

4.7.2.3 External Hoop Stress (f_h)

Thin walled tubular members, such as steel pipe, will fail by buckling or collapse when under the influence of a sufficient external hoop stress. A traditional formula established by Timoshenko for calculation of the wall thickness required to prevent collapse of a round steel pipe is as follows (13).

$$t = D/12(864P_{ext}/E)^{1/3}$$

Where:

P_{ext} = uniform external pressure in pounds per square inch

Since pipe in an HDD pull section will not necessarily be perfectly round and will be subject to bending and dynamic loading, a conservative factor of safety should be applied in checking pipe wall thickness using the above relationship. Generally speaking, diameter to wall thickness ratios for steel pipe to be installed by HDD should be held at 60 or below although higher D/t ratios are appropriate if a high level of confidence exists in collapse analysis calculations or a counterbalancing internal pressure will be applied during pullback (14).

As with bending, hoop stress resulting from external pressure can be checked using criteria established for tubular members in offshore structures (6). Applicable formulas are presented below (2).

$f_h = P_{ext}D/2t$

$F_{he} = 0.88E(t/D)^2$ for long unstiffened cylinders

$F_{hc} = F_{he}$ for $F_{he} \leq 0.55F_y$

$F_{hc} = 0.45F_y + 0.18F_{he}$ for $0.55F_y < F_{he} \leq 1.6F_y$

$F_{hc} = 1.31F_y/(1.15 + (F_y/F_{he}))$ for $1.6F_y < F_{he} \leq 6.2F_y$

$F_{hc} = F_y$ for $F_{he} > 6.2F_y$

Where:

f_h = hoop stress due to external pressure in pounds per square inch
F_{he} = elastic hoop buckling stress in pounds per square inch
F_{hc} = critical hoop buckling stress in pounds per square inch

155

Using these formulas, hoop stress due to external pressure should be limited to 67% of the critical hoop buckling stress.

4.7.2.4 Combined Installation Stresses

The worst-case stress condition for the pipe will typically be located where the most serious combination of tensile, bending, and external hoop stresses occurs simultaneously. This is not always obvious in looking at a profile of the drilled hole because the interaction of the three loading conditions is not necessarily intuitive. To be sure that the point with the worst-case condition is isolated, it may be necessary to perform a combined stress analysis for several suspect locations. In general, the highest stresses will occur at locations of tight radius bending, high tension (closer to the rig), and high hydrostatic head (deepest point) (6).

Combined stress analysis may begin with a check of axial tension and bending according to the following limiting criterion (6). The criterion is taken from practices established for design of tubular members in offshore structures with an increase in the allowable tensile proportion to make it consistent with established practice in the HDD industry (2).

$$f_t/0.9F_y + f_b/F_b \leq 1$$

Where:

f_t = tensile stress in pounds per square inch

The full interaction of axial tension, bending and external pressure stresses should be limited according to the following criteria (2) (6).

$$A^2 + B^2 + 2v|A|B \leq 1$$

Where:

A = $((f_t + f_b - 0.5f_h)1.25)/F_y$

B = $1.5f_h/F_{hc}$

v = Poisson's ratio, 0.3 for steel (4)

It should be noted that failure to satisfy the unity checks presented above does not mean that the pipeline will necessarily fail by overstress or buckling. Rather, it indicates that the combined stress state places the design in a range where some test specimens under similar stress states have been found to be subject to failure (6).

4.7.3 Operating Loads

As with a pipeline installed by conventional methods, a pipeline installed by HDD will be subjected to internal pressure, thermal expansion, and external pressure during normal operation. However, a continually welded or fused pipeline installed by HDD will also be subjected to elastic bending. The operating loads imposed on a pipeline installed by HDD are described below.

4.7.3.1 Internal Pressure

A pipeline installed by HDD is subjected to internal pressure from the fluid flowing through it. For design purposes, this pressure is generally taken to be the pipeline's maximum allowable operating pressure. The internal hydrostatic pressure from the depth of the HDD installation should be considered when determining the maximum internal pressure.

4.7.3.2 Bending

Elastic bends introduced during pullback will remain in the pipe following installation and therefore must be considered when analyzing operating stresses. These bends are typically approximated as circular curves having a radius of curvature that is determined from as-built pilot hole data. One common method of calculating the radius of an approximate circular curve from pilot hole data is presented below (6).

$$R = (L/A)688$$

Where:

R = radius of curvature of the drilled hole in inches

L = length drilled in feet

A = the total change in angle over L in degrees

The selection of a value for L is based on engineering judgment and takes into account the actual curvature of the pipe installed in the reamed hole as opposed to individual pilot hole survey deflections.

4.7.3.3 Thermal Expansion

A pipeline installed by HDD is considered to be fully restrained by the surrounding soil. Therefore, stress will be induced by a change in temperature from that existing when the line was constructed to that present during operation.

It should be noted that the fully restrained model is not necessarily true for all subsurface conditions. Obviously, a pipeline is not fully restrained during installation or it could not be pulled through the hole. Engineering judgment must be used in considering thermal stresses and strains involved with an HDD installation.

4.7.3.4 External Pressure

In order to evaluate the impact of external pressure during operation, the minimum internal operating pressure of the pipeline should be compared against the maximum external pressure resulting from groundwater and earth load at the lowest elevation of the HDD installation.

The earth load on pipelines installed by HDD is generally a "tunnel load", where the resulting soil pressure is less than the geostatic stress. ASTM F 1962-99 recommends the following method for calculating earth loads on HDD installations (5).

$$P_e = \kappa \gamma H / 144$$

Where:

P_e = external earth pressure in pounds per square inch

κ = arching factor

γ = soil weight in pounds per cubic foot

H = depth of cover in feet

The arching factor is calculated as follows.

$$\kappa = (1-\exp((-2KH/B)\tan(\delta/2)))/((2KH/B)\tan(\delta/2))$$

Where:

K = earth pressure coefficient

B = "silo" width in feet which is assumed to be the reamed hole diameter

δ = angle of wall friction in degrees which is assumed to equal ϕ

ϕ = soil internal angle of friction

The earth pressure coefficient is calculated as follows.

$$K = \tan^2(45-\phi/2)$$

4.7.4 Operating Stresses

With one exception, the operating stresses in a pipeline installed by HDD are not materially different from those experienced by pipelines installed by cut and cover techniques. As a result, past procedures for calculating and limiting stresses can be applied. However, unlike a cut and cover installation in which the pipe is bent to conform to the ditch, a pipeline installed by HDD will contain elastic bends. Bending stresses imposed by the HDD installation process should be checked in combination with other operating stresses to evaluate if acceptable limits are exceeded. Other longitudinal and hoop stresses that should be considered will result from internal pressure and thermal expansion/contraction (6).

4.7.4.1 Internal Hoop Stress (f_h)

Hoop stress due to internal pressure is calculated as follows (4).

$$f_h = (P_{int}D)/(2t)$$

Where:

f_h = hoop stress due to internal pressure in pounds per square inch

P_{int} = uniform internal pressure in pounds per square inch

The maximum allowable hoop stress due to internal pressure will be governed by the design standard applicable to the pipeline transportation system that contains the HDD segment being examined. For example, hoop stress is limited to 72% of the specified minimum yield strength for liquid petroleum pipelines (4). For natural gas pipelines, hoop stress limitations range from 40% to 72% of the specified minimum yield strength (8).

4.7.4.2 Bending Stress (f_b)

Bending stresses are calculated and limited as shown in Section 4.7.2.2.

4.7.4.3 Thermal Stress (f_e)

Thermal stress resulting from changes in pipe temperature from the point in time at which the pipe is restrained by the surrounding soil to typical operating condition is calculated as follows (4).

$$f_e = E\alpha(T_2-T_1)$$

Where:

f_e = longitudinal stress from temperature change in pounds per square inch

α = coefficient of thermal expansion for steel in inches per inch per °F

T_1 = temperature at installation, or when the pipeline becomes restrained, in °F

T_2 = operating temperature in °F

The high thermal conductivity of steel enables the temperature of the pipe to equalize with the surrounding soil within a matter of hours after construction. Since soil temperatures at the depth of most HDD installations are relatively constant, thermal stresses are typically a concern only when the temperature of the product flowing through the pipeline differs substantially from that of the

159

surrounding soil, such as in hot oil pipelines or immediately downstream of a natural gas pipeline compressor station.

4.7.4.4 Combined Operating Stresses

Hoop, thermal, and bending stresses imposed on the pipe during operation should be checked in combination to evaluate the risk of failure from combined stresses. This can be accomplished by examining the maximum shear stress at selected elements on the pipe. Maximum shear stress is calculated by the following formula (17).

$$f_v = (f_c - f_l)/2$$

Where:

f_v = maximum shear stress in pounds per square inch

f_c = total circumferential stress in pounds per square inch

f_l = total longitudinal stress in pounds per square inch

In this analysis, all tensile stresses are positive and compressive stresses are negative. The total circumferential stress is the difference between the hoop stress due to external pressure and the hoop stress due to internal pressure. The total longitudinal stress is the sum of the bending and thermal stresses and the longitudinal component of circumferential stress determined as follows.

$$f_{lh} = f_c \nu$$

Where:

f_{lh} = longitudinal component of circumferential stress in pounds per square inch

Presuming that hoop stresses will be positive for pressurized steel pipelines, the pipe element that will typically have the highest maximum shear stress is that which has the highest total longitudinal compressive stress. This element will fall the maximum distance from the neutral axis on the compression side of an elastic bend. Maximum shear stress should be limited to 45% of the specified minimum yield strength (4).

4.7.5 Pulling Load Calculation by the PRCI Method

Calculation of the approximate tensile load required to install a pipeline by HDD is relatively complicated due to the fact that the geometry of the drilled path must be considered along with properties of the pipe being installed, subsurface materials, and drilling fluid. Assumptions and simplifications are required. A method to accomplish this was published by the Pipeline Research Committee at the American Gas Association, now known as the Pipeline Research Council

160

International (PRCI), in the 1995 Edition of *Installation of Pipelines By Horizontal Directional Drilling*.

The PRCI Method involves modeling the drilled path as a series of segments to define its shape and properties during installation. The individual loads acting on each segment are then resolved to determine a resultant tensile load for each segment. The estimated force required to install the entire pull section in the reamed hole is equal to the sum of the tensile loads acting on all of the defined segments.

In utilizing the PRCI Method, engineers should be aware that pulling loads are affected by numerous variables, many of which are dependent upon site-specific conditions and individual contractor practices. These include prereaming diameter, hole stability, removal of cuttings, soil and rock properties, drilling fluid properties, and the effectiveness of buoyancy control measures. Such variables cannot easily be accounted for in a theoretical calculation method designed for use over a broad range of applications. For this reason, theoretical calculations are of limited benefit unless combined with engineering judgment derived from experience in HDD construction.

4.7.5.1 Drilled Path Model

Drilled path models can be based on the designed drilled path, a "worst-case" drilled path, or "as-built" pilot hole data, if available. Bearing in mind that most pilot holes are drilled longer, deeper, and to tighter radii than designed, a conservative approach in the absence of as-built pilot hole data is to evaluate a worst-case drilled path which takes into account potential deviations from the design. This worst-case path should be determined based on allowable tolerances for pilot hole length, elevation, and curve radius. Significant deviations in these parameters are typical and often result from conditions beyond the control of drilling contractors. For example, it would not be unusual to find deflections in a pilot hole that produce a bending radius approaching 50% of the design radius.

A Cartesian coordinate system must be established for the model. Typically, the coordinate system will originate at the entry point. A model segment is described by its true length and the inclination and azimuth angles at its endpoints.

Using this information, the relative positions of the segment endpoints and radius of curvature, if the segment is curved, are calculated using the "Minimum Curvature Method" (3). This information is then used to calculate the installation loads acting on the segment.

4.7.5.2 Straight Segment Loads

As previously discussed in Section 4.7.1.1, tension on the segment results from frictional drag between the pipe and the wall of the hole, fluidic drag from viscous drilling fluid surrounding the pipe, and the effective weight of the pipe.

4.7.5.3 Curved Segment Loads

Pulling loads resulting from fluidic drag and effective weight are calculated in the same manner as a straight segment. The angle used for resolving the effective weight into an axial component is the average of the angles at the ends of the segment.

Calculation of the pulling load resulting from frictional drag is more complicated. Radial forces involved with bending must be taken into account. The PRCI Method does this by modeling a curved segment as a beam in three point bending. The curvature and deflection of the beam are known from the data input in the drilled path model. The normal force can then be calculated as follows (16).

$$N = (Th-Wcos\theta(Y/144))/(X/12)$$

$$h = R(1-cos(\alpha/2))$$

$$Y = 18L^2-j^2(1-1/cosh(U/2))$$

$$U = 12L/j$$

$$j = (EI/T)^{\frac{1}{2}}$$

$$X = 3L-(j/2)tanh(U/2)$$

Where:

N = Normal force in pounds
T = Average tension over the segment, $(T_1+T_2)/2$), in pounds
h = Center displacement of the segment in feet
R = Radius of curvature of the segment in feet
α = Deflection angle (dogleg angle) of the segment
W = Weight of the segment that contributes to bending expressed in pounds
θ = Angle of the segment relative to horizontal
L = Segment length in feet

The pulling load resulting from the normal force is calculated by multiplying the normal force and associated reactions (N/2) by the coefficient of friction. Since a value of T, the average tension, must be assumed to calculate the normal force, an iterative calculation is required to arrive at the correct value.

If the bend is in the horizontal plane, the normal force is calculated without a component of segment weight since the weight is not in the plane of the bend.

4.7.6 External Coating

External coatings used in HDD installations should be smooth and resistant to abrasion. Historically, pipelines installed by HDD in alluvial soils have been coated with corrosion coating only. Weight coating is generally not required. The deep, undisturbed cover provided by HDD installation has proven adequate to restrain buoyant pipelines.

The corrosion coating most often used on HDD crossings is thin film fusion bonded epoxy. This coating is popular not only because it is highly durable, but also because the field joints can be coated using a compatible fusion bonded epoxy system. For crossings installed in rock, highly abrasive soils, or soil conditions that might involve point loads, a protective coating should be used in addition to the corrosion coating. The protective coating doesn't need to have corrosion prevention properties; it should simply protect the underlying corrosion coating.

4.8 Contractual Considerations

Once design of the crossing is complete, a set of contract documents should be produced for solicitation of bids from contractors and to govern construction of the crossing. Contract documents should be structured to clearly present technical, commercial, and legal requirements. Contract forms applied to HDD projects can be separated into two basic categories: *lump sum* and *daywork*. A lump sum contract is one in which the contractor is paid a fixed amount for delivering a drilled segment in accordance with plans and specifications. Payment is based on performance and does not vary with the time or effort expended. A daywork contract is one in which the contractor is paid a fixed amount per day, or some other unit of time, for providing a spread of equipment in accordance with the contract documents.

4.8.1 Lump Sum Contracts

In most cases, HDD installations should be bid using standard lump sum contract forms. Applicable technical specifications and drawings should be included in the contract documents. Bid prices may be broken down for convenience or analysis; however, compensation should generally be on a lump sum basis as opposed to a fixed unit price. For example, confusion may result on a crossing priced on a per foot basis if drilling conditions dictate that a slightly longer drilled length is easier to complete than the designed drilled length. It is not reasonable for a contractor to extend the length of a crossing solely for his convenience, and in so doing increase his compensation. It is also unreasonable for the owner to insist that the crossing be redrilled to a shorter length to reduce payment. Setting a lump sum

163

price for a crossing installed in accordance with the plans and specifications eliminates these problems.

4.8.2 Specifications and Drawings

An HDD technical specification should clearly define all details relative to HDD performance. Job specific details such as pilot hole tolerances, water sources, and drilling fluid disposal requirements should be included in the specification or an accompanying job-specific section. In addition to the technical specification, the contract documents should contain a plan & profile drawing. The drawing should complement the technical specification by providing a clear presentation of the crossing design as well as the results of topographic, hydrographic, and geotechnical surveys.

4.8.3 Daywork Contracts

Because of the evolving nature of HDD technology, the industry has employed many contract form variations. Typically these variations involve negotiating some type of completion incentive into a daywork contract. Significant technical advances have been made because of owner willingness to assume the risk of cost overruns or completion failure for prospective HDD installations that were not contractually feasible. However, a daywork contract requires much greater oversight by the owner than a typical lump sum contract.

Although a daywork contract may not specify contractor performance in terms of a completed installation, contractor performance is required and should be clearly defined. The required performance primarily involves the provision of equipment of a certain capacity. The components of equipment should be listed as well as conditions with respect to downtime, maintenance, crewing, fueling, etc. Additionally, items in the scope of work that can be contracted on a lump sum basis, such as mobilization and site preparation, should be broken out, priced lump sum, and governed by an appropriate performance specification.

4.9 Construction Monitoring

The primary objectives of an inspector involved in construction monitoring on an HDD installation are to assist in the interpretation of the contract documents and to document conformance, or non-conformance, by the drilling contractor. In doing this, it is important for the inspector to document his observations and actions. Should a question or dispute arise after the installation is complete, the inspector's notes may provide the only source of confirming data. Since a drilled installation is typically buried with deep cover under an inaccessible obstacle, its installed condition cannot be confirmed by visual examination.

164

4.9.1 Directional Performance

The inspector should be concerned with directional drilling performance in two basic areas: position and curvature. First, the contractor must install the pipeline so that the drilled length and depth of cover specified by the contract are provided. Second, the contractor must not curve the drilled path in such a way that the pipeline will be damaged during installation or overstressed during operation. The actual position of the drilled path cannot readily be confirmed by an independent survey. Therefore, it is necessary for the inspector to have a basic understanding of the downhole survey system being used by the contractor and be able to interpret its readings. It is not necessary for the inspector to observe and approve the drilling of each joint, however, progress should be monitored on a daily basis and problems addressed so that remedial action can be taken as soon as possible. The inspector should ensure that bends are not drilled at a radius of curvature less than the minimum allowable. If a tight radius occurs, the unacceptable portion of the hole should be redrilled. If redrilling proves unsuccessful, the tight radius should be reviewed with the design engineers to ensure that the codes and specifications governing design of the pipeline are not violated.

4.9.2 Drilling Fluid

The inspector should document all drilling fluid products being used, the contractor's pumping pressures and rates, and details relative to drilling fluid circulation at the HDD endpoints. The right-of-way and surrounding areas should be examined regularly for inadvertent returns. If inadvertent returns are discovered, they should be contained or cleaned up in accordance with the specifications and permits and their locations should be monitored for continuing problems.

4.9.3 Additional Concerns

Depending upon the contractual and technical details of an HDD installation, there are numerous details that should be reviewed and documented. These may consist of gauge readings, production rates, equipment failure, downtime, etc. During pullback, it is important to document the contractor's operations relative to pull section handling. Additionally, buoyancy control measures, if used, should be documented. Following the completion of pullback, the condition of any visible pipe and coating at the leading edge of the pull section should be documented.

References

(1)   American Gas Association, Drilling Fluids in Pipeline Installation By Horizontal Directional Drilling, A Practical Applications Manual, Washington: Pipeline Research Committee at the American Gas Association, 1994, 30.

(2)   ANSI/API RP 2A-WSD-93, Recommended Practice for Planning, Designing and Constructing Fixed Offshore Platforms – Working Stress Design, Twentieth Edition, Washington: American Petroleum Institute, 1993, 40-44.

(3)   API Bulletin D20, Directional Drilling Survey Calculation Methods and Terminology, 1985.

(4)   ASME/ANSI B31.4-1986 Edition, Liquid Transportation Systems for Hydrocarbons, Liquid Petroleum Gas, Anhydrous Ammonia, and Alcohols, New York: American Society of Mechanical Engineers, 1986, 9-28.

(5)   ASTM Designation: F 1962-99, Standard Guide for Use of Maxi-Horizontal Directional Drilling for Placement of Polyethylene Pipe or Conduit Under Obstacles, Including River Crossings, West Conshohocken, PA: American Society for Testing Materials, 1999, 15.

(6)   American Gas Association, Installation of Pipelines By Horizontal Directional Drilling, An Engineering Design Guide, Washington: Pipeline Research Committee at the American Gas Association, 1995, 41-80.

(7)   Bleier, Roger et al., "Drilling Fluids: Making Peace with the Environment", Journal of Petroleum Technology, vol. 45, no. 1, January 1993, 6-10.

(8)   Code of Federal Regulations, Title 49, Part 192, Transportation of Natural and Other Gas by Pipeline: Minimum Federal Standards, 2001, 192.111.

(9)   Fowler, J.R. and Langner, C.G., "Performance Limits for Deepwater Pipelines", OTC 6757, 23rd Annual Offshore Technology Conference, Houston, TX, May 6-9, 1991.

(10)   HDD Consortium, Horizontal Directional Drilling, Good Practices Guidelines, 2001, 3-25.

(11)   Loh, J.T., "A Unified Design Procedure for Tubular Members", OTC 6310, 22nd Annual Offshore Technology Conference, Houston, TX, May 7-10, 1990.

(12)   Lummus, James L. and Azar, J.J., Drilling Fluids Optimization, A Practical Field Approach, Tulsa, Oklahoma; PennWell Publishing Company, 1986, 111-112.

(13)   Merritt, Frederick S., Standard Handbook for Civil Engineers, New York: McGraw-Hill, Inc., 1968, 21-37.

(14) O'Donnell, Hugh W., "Investigation Of Pipeline Failure In A Horizontally Directionally Drilled Installation", Pipeline Crossings 1996, Proceedings of Specialty Conference, Pipeline Division ASCE, 1996.

(15) Puckett, Jeffrey S., "Analysis of Theoretical Versus Actual HDD Pulling Loads", Volume Two, New Pipeline Technologies, Security and Safety, Proceedings of the ASCE International Conference on Pipeline Engineering and Construction sponsored by The Technical Committee on Trenchless Installation of Pipelines (TIPS) of the Pipeline Division of ASCE, Baltimore, Maryland, July 13-16, 2003, 1352.

(16) Roark, R.J., Formulas for Stress and Strain, Second Edition & Fifth Edition, New York, New York; McGraw-Hill, 1943, 1965

(17) Timoshenko, S. P. and Gere, James M., Mechanics of Materials, New York, New York; Van Nostrand Reinhold Company, 1972, 9-48.

(18) Young, Warren C., Roark's Formulas for Stress and Strain, Sixth Edition, New York: McGraw-Hill, Inc., 1989, 94-95.

5. BUOYANCY

5.1 Introduction

This chapter presents design and construction information on providing buoyancy control for overland pipelines. Pipelines are subject to natural positive buoyancy when placed below the water table. The magnitude of the buoyant effect is proportional to the pipeline diameter. The need to mitigate the pipe buoyancy depends on many factors. These factors include the pipeline material, the pipe diameter and its wall thickness, the pipe coating, the pipe product, the time of installation, and the density and type of trench backfill soil or imported material. Even pipelines carrying a heavy product may require the use of buoyancy control measures as they could be subject to excessive upward buoyancy forces in the period between pipe installation and full operation. All factors should be taken into account when determining the optimum type and frequency of buoyancy control measures.

Buoyancy control is rarely a pipe integrity issue. The floating of a pipe to the ground surface can relieve pipe stress rather than add to it. However, pipe exposure can result in costly repairs to rebury the pipe, and can expose the pipe to other dangers, such as third party damage or damage from debris carried by fast flowing watercourses. Thus, it is usually desirable to maintain a minimum soil cover over cross-country pipelines.

This chapter describes various buoyancy design criteria as well as typical, or industry standard, design values. A design process is presented along with representative calculations. It also presents many of the various pipeline buoyancy control methods available to designers and contractors. The methods include those that are more the typical industry standards, as well as other less well known methods. The latter methods, although less well known, are acceptable buoyancy control measures.

Thermal expansion forces and overbend stability, although important considerations in keeping a buried pipe below the ground surface, are not part of this section.

Also presented are various references to past ASME papers on buoyancy control design, products, and field implementation.

5.2 Definitions and Abbreviations

ASTM — American Standards Testing Materials

CCC — Continuous Concrete Coating, a concrete layer of specified thickness that encircles the pipe and provides buoyancy control in watercourses, canals, or water-filled pipeline trenches

CSA — Canadian Standards Association

DP — Pipeline Design Pressure

Ditch Plug — See Trench Plug

Diversion Berm — An approximately 1 m high by 1 m wide mound of mineral soil placed across the RoW at 10-60 m spacing along the RoW used to divert surface water off the pipeline right-of-way and into vegetated terrain.

FA — Frost or Slurry Anchor, steel rods frozen into permafrost soil and attached to a saddle draped over the pipeline to provide buoyancy control

GA — Grouted Anchor, steel rods grouted into the lower competent mineral soil and attached to a saddle draped over the pipeline to provide buoyancy control

GSC — Geological Survey of Canada

GSW — Geotextile Swamp Weights, a geotextile fabric draped over the pipeline and up the trench walls and backfilled with mineral soil to a minimum specified thickness to provide buoyancy control

Gouge Auger — A small (25 mm) diameter approximately 1.8 m long metal rod with a 300 mm long hollow base section used to probe muskeg depths in unfrozen ground

Lagging — Wood boards placed between river weights and usually banded with metal strips

Muskeg — Another name for high organic content, wet terrain. (See Organic Terrain)

NPS — Nominal Pipe Size

1-D Terrain Typing — A takeoff of terrain types on a linear basis along the proposed pipeline centreline (Please refer to the chapter on terrain analysis.)

Organic Terrain — Areas that contain thick, often defined for pipelines as greater than 300 mm, of high organic content (fibrous) soils, or peaty soils, and that generally have a near surface water table.

PRC — Pipeline Research Committee

Permafrost	Any rock or soil material that has remained below 0 °C continuously for two or more years. The two year stipulation is meant to exclude from the definition the overlying ground surface layer which freezes every winter and thaws every summer (seasonal frost).
	Cold permafrost - Permafrost which has an average yearly temperature below -2 °C (this number will vary depending upon the specific engineering application under consideration). Cold permafrost may tolerate the introduction of considerable heat without thawing.
	Warm permafrost - Permafrost with an average yearly temperature between 0 °C and -2 °C. The addition of small amounts of heat may induce thawing.
	Discontinuous permafrost - An area in which there co-exist pockets of permafrost and unfrozen soil. The percentage of each varies dependant upon the location.
	Active Layer – The upper ground surface layer which, thaws every summer and re-freezes every winter in permafrost areas.
Phi Angle	Internal soil or muskeg friction angle; and for the purposes of this chapter - mainly related to organics or organic/mineral soil mixes
PipeSak™ Weight*	Prefabricated granular filled bags with lower 'baffles' that are set on to the pipeline to provide buoyancy control
PW	Plate Weight, a solid high density thin plastic 'hat' placed over the pipeline, which extends laterally out from the pipeline, and is backfilled with mineral soil to a minimum thickness to provide buoyancy control
Roach	The mound of soil over the backfilled pipeline which protrudes above the adjacent grade at the end of pipeline construction
RoW	Right-of-way
RW	River Weight or "bolt on" encirclement type concrete weight that provides buoyancy control in very wet trenches
SA	Screw Anchor, steel rods typically with a single helix embedded in competent mineral soil and attached to a saddle draped over the pipe that provides buoyancy control
SMYS	Specified Minimum Yield Strength
Soil Density	The unit weight of soil in kg/m3 (total or wet unit weight unless otherwise specified)

Stub Berm	A short diversion berm, approximately 4 m by 2 m by 1m high, made of mineral soil placed at 30 -100 m spacing along the RoW centered over the backfilled trench, usually in areas of muskeg with a gentle slope.
SW	Swamp Weight or saddle weight or "set on" type concrete weight that provides buoyancy control in drier trenches.
Terrain Type	A representation of the surface and near surface soil types on the basis of geological origin, landform, and texture. (Please refer to the chapter on terrain analysis.)
Thaw sensitivity	A property of the permafrost, which is dependant upon both the amount and disposition of the ice within the soil.
Thaw stable permafrost	Permafrost in which the ice within the soil is interstitial and confined to the pores. Subsidence or settlement when thawed is minor, and the foundation remains essentially sound.
Thaw sensitive permafrost	Permafrost wherein the ice within the soil is present either in a segregated form, or occurs outside of the pore structure. The result of thawing can be the loss of strength, excessive settlement and soil containing so much moisture that it flows.
Trench Plug	A full trench width barrier of polyurethane foam or powdered bentonite bags, or sand bags, placed prior to backfilling, around and over the pipeline and keyed into the trench to block water flow along the trench.
WT	Pipeline Wall Thickness

* Registered Trademark of the PipeSak Inc. Company of Ontario

5.3 Pipeline Codes

Various pipeline standards, such as the ASME B31 volumes in the United States and the Canadian pipeline standard, CSA Z662, have clauses requiring buried pipelines to have a minimum depth of cover. The required depth of cover varies with the terrain crossed; but is typically a minimum of 0.6 metres (m) to 0.9 m within overland muskeg areas and 1.2 m below channel bottom at watercourses.

When a pipeline is placed in an area with lightweight soils (organics) in combination with a potential water table near grade (muskeg) and the pipeline has a buoyant force that exceeds its self weight plus the resistance of the overburden, the pipeline will float. To maintain the minimum required depth of cover under these conditions, buoyancy control measures are required.

5.4 Buoyancy Design Philosophy

The assessment effort or 'engineering' process used when quantifying or locating buoyancy control on pipelines varies widely. The philosophy may range from simply placing buoyancy control over 100% of the pipeline because it's relatively inexpensive and easy, to a detailed desktop study with a subsequent detailed field investigation program(s) that optimizes buoyancy control so that it is placed only where it's needed. The latter approach requires an appropriate level of understanding of the factors involved, and an agreed upon (between design, construction and operations) criteria that balances cost vs. operational risk. The level of effort should reflect the diameter/length combination of the pipeline. A high level of effort may not be warranted for shorter, small diameter lines, but could well save thousands and even millions of dollars for long, large diameter pipelines (Robertson).

There are two main areas where buoyancy control is applied to overland pipelines:

1. Swamps or Muskeg areas (deep (relative to the pipe diameter), low weight, highly organic soils with a high water table)

2. Watercourse crossings or waterways (the latter referring to canals or rivers used as the RoW for the pipeline)

A third situation where buoyancy control may be considered is where soils are subject to liquefaction during an earthquake. Fluidization of a soil can lead to very high buoyancy forces. Depending on the product in the pipeline, these high buoyancy forces may exceed the pipeline's overall mass. The decision to apply buoyancy control in these areas must also consider the possible strain relief gained by a pipeline rising to the surface and partly avoiding the significant forces imposed by a large moving soil mass.

Buoyancy control is not necessarily required in all areas that have a high water table. If the soils are not susceptible to liquefaction and have suitable weight and strength, buoyancy control is not required.

173

5.5 Buoyancy Control Options

There are many different types of buoyancy control options available to the designer. These include:
- concrete saddle weights
- concrete bolt-on weights
- continuous concrete coating
- deeper ditch (additional mineral soil cover)
- soil filled "bag" weights
- soil filled geotextile swamp weights
- plate weights
- imported fill
- steel screw anchors
- grouted steel anchors
- "freeze-back" steel anchors in continuous permafrost areas

Each weight type is discussed in more detail in the following sections.

5.5.1 Concrete Weights

The three common types of concrete weighting are:
- saddle weights, sometimes referred to as swamp weights
- continuous concrete coating
- bolt-on weights, sometimes referred to as river weights

5.5.1.1 Saddle Weights

Concrete saddle weights (SW), also known as swamp weights, are a common weighting option for drier organic terrain (muskegs). Swamp weights are used in areas where the pipeline can be welded beside the trench and lowered-in using conventional land lay methods, or in terrain where the trench either does not fill up with water or can be dewatered with minimal effort. If the trench cannot be pumped dry, river weights or continuous concrete coating are used. Swamp weights are lowered onto the pipeline after it has been lowered into the trench. (See Figures 1 and 2) The inside of the weights will usually have a felt coating to protect the pipe coating from damage.

The trench usually must be dug wider and deeper in areas where swamp weights are used in order to accommodate the swamp weight's width and height. Typically, minimum pipe cover is specified to be applied from the top of the swamp weight, not from the top of pipe between the weights. The increased ditch volume may be in the order of 35% to 50%.

174

Swamp weights typically are produced at a central manufacturing facility. They may be produced at a temporary remote batch plant near the pipeline project if the costs can be justified based on savings on shipping of the weights. A suitable local concrete quality aggregate supply is usually needed to justify a remote batch plant.

Swamp weights are not typically used within watercourse channels. They may, however, be placed at the ends of a watercourse section as additional buoyancy protection if the continuous concrete coating does not extend to or slightly beyond the edge of water.

Figure 1: Typical Concrete Saddle Weight (NPS 42)

Figure 2: Typical Installed Concrete Saddle Weight

5.5.1.2   Continuous Concrete Coating

Continuous concrete coating, or CCC, is commonly used on larger watercourse crossings. It is also used in organic areas or low weight backfill areas where rapid water inflow into the trench cannot be effectively removed. CCC also may be used in areas where the backfill soils are subject to liquefaction (assuming, from a strain relief viewpoint, that it isn't preferable to allow the pipe to come to the surface.)

The pipe trench may need to dug larger than normal to accommodate concrete coated pipe, but does not need to be as large as for either swamp weights or river weights.

CCC can be applied on site or it can be pre-applied in a shop. Shop coated pipe usually results in a better quality coating due to improved curing conditions. Shop coating also eliminates the need for on-site hoarding and heating for winter construction projects. However, the coated pipe sections result in higher shipping costs and a reduced number of pipe joints per haul truck because of the increased weight and volume. For shop coated pipe, the field joints may be concrete coated, or may be left uncoated. The designer may take the conservative approach and assume that field joints are not coated. Joints of concrete coated pipe are shown in Figure 3.

176

For onsite applications, one technique is to pour wet concrete into pipeline diameter specific forms placed around the pipeline. In winter, the concrete will have to be hoarded and heated as per appropriate construction specifications in order to cure properly and attain its required properties, such as minimum compressive strength.

A second technique is to spray apply concrete to the pipe. Proper application techniques are required to prevent damage to the pipe coating.

CCC at watercourse crossings allows the weighted section to be dragged or floated (see Figure 4) into place within the channel. This is an advantage over river weights, which must be lifted and lowered into place. CCC has the added advantage of providing continuous physical protection to the pipeline against external damage from things such as rocks in the backfill.

Figure 3: Typical Concrete Coated Pipe (courtesy Garrie Cox)

Figure 4: Floating "Push" Section of Concrete Coated Pipe (courtesy Garrie Cox)

5.5.1.3   River Weights

River weights (RW), or sometimes called bolt-on weights, are typically used in smaller watercourses, or in organic terrain that cannot be dewatered. River weights may be preferred over CCC for smaller pipe diameters and/or shorter weighted lengths. River weight halves bolted together are attached to each side of the pipeline prior to it being lowered into the trench. (See Figures 5 and 6)The inside of the weights will usually have a felt coating to protect the pipe coating from damage.

Similar to swamp weights, the pipe trench usually must be dug wider and deeper to accommodate the weights and to meet minimum cover requirements.

River weights, like swamp weights, are typically produced at an established manufacturing plant. They may, for project specific reasons and conditions, be produced at a remote batch plant.

Figure 5: Individual River Weight Halves

Wood lagging, see Figure 7, is generally attached to the pipeline between river weights used at watercourses. The lagging ensures the proper weight spacing is maintained during lowering-in of the pipe, and also provides mechanical protection to the pipe. Lagging typically consists of either "one by fours" or "two by fours" banded together and wrapped around the pipeline. Lagging may be omitted if, at the discretion of the field supervisor, it is considered unnecessary. It may be omitted when RWs are used in swamps as the risk of damage to the pipe from backfill is very low, and because the pipe is simply lowered into the trench vs. being carried some distance from a work area to the watercourse.

Figure 6: River Weighted Pipe Section Ready for Lowering-in (courtesy John Ness of World Wide CMS)

Figure 7: River Weighted Pipe Section with Wood Lagging

5.5.2   Concrete Weight Dimensions

The dimensions of saddle and river weights vary from supplier to supplier. Many suppliers have pre-made forms available. Weights may be cheaper to make using a manufacturer's standard forms vs. making project specific forms for a particular pipeline project. Larger pipeline companies may have their own specific requirements for weights sizes and mass and based on volumes can get their weights produced cost effectively.

The final swamp or river weight dimension and mass should take into consideration:

- constructability, both in the building of the weights and in the field placement
- saddle weight stability in the possibly soft trench bottom
- typical construction equipment for a particular pipe diameter
- reasonable weight spacing
- pipe stress due to the mass of the weight (considering full weight mass in overland situations where the water table may drop below the pipe invert)

Thus, although weights from various suppliers may vary in size and mass for a particular pipe diameter, they are generally similar. Table 1 provides an indication of the typical mass and dimensions for saddle and bolt-on weights for a few pipeline diameters. Pipe stress is usually not a problem when using a supplier's standard weights. In any case, all weights should be made to approved engineered drawings using approved concrete mixes and other appropriate concrete standards.

Pipe Diameter	Saddle and Bolt-On Weight Mass (kg)	Saddle Weight Length and Height (mm)	Bolt-On Weight Length and Thickness (mm)
NPS 6	200	670 x 340	600 x 300
NPS 12	700	1000 x 550	1000 x 450
NPS 24	2400	1500 x 900	1400 x 600
NPS 36	3600	1500 x 1200	1600 x 600
NPS 48	6400	1800 x 1600	1800 x 600

Table 1: Typical Approximate Saddle and Bolt-On Weight Mass

5.5.3　Concrete Reinforcement

Steel reinforcement is required for all concrete weights. It may be as simple as a narrow gauge wire mesh centered in the CCC for a small diameter pipeline. Swamp or river weights for a large diameter pipe will require a significant amount of reinforcing steel. Weight reinforcement needs to be properly designed and the design approved by a professional engineer.

5.5.4　Soil Weights

Soil weights use either native soil or imported soil for providing additional mass over the pipe. These weight types are used almost exclusively for overland buoyancy control vs. at watercourses. The two main types of soil weighting are:

- using in-situ soil by digging a deeper trench
- bag weights filled with granular soil

Two other soil type weights that have been tested and used on pipelines in Alberta, Canada are:

- geotextile swamp weights
- plate weights

Each option is further described in the following sections.

Soil weights, depending on the fill material, generally have less mass per unit volume than concrete weights. However, their submerged mass may exceed that of concrete due to the increased saturation of a high density, coarse grained granular soil vs. non-porous concrete. But, the main advantage of soil weighing is its flexibility. During construction, the weights can be easily moved to where they are required. The weights usually have lower preconstruction capital costs and thus more weights than may be necessary can be ordered without a large cost premium. Unused, unfilled weights can be shipped back to a warehouse or storage facility without incurring high transportation costs.

Various soil weight types have been patented and sourcing may be restricted to certain vendors.

5.5.4.1　Deeper Trench

Deepening the pipe trench is the most cost effective way of providing buoyancy control. It is used where lowering the pipeline trench bottom an additional 300 mm to 600 mm results in the excavation of sufficient additional in situ higher density mineral soil, plus additional shear force, to offset the pipe's natural buoyancy. The cost of the additional trenching is usually well below the combined supply and install cost of any manufactured buoyancy control measure.

One potential disadvantage of using a deeper trench is increased post construction settlement of the ditch backfill. The increased ditch depth leads to greater relative backfill consolidation and settlement. Backfill settlement can be more severe in areas of organic soils due to the destruction of the "structure" of the organic mat during trenching. In northern areas where pipelines often are constructed in the winter due to improved access over the frozen ground, the higher ice content and often "blocky" nature of the excavated/backfilled soil can also add to post construction settlement. This may result in a swale developing along the ditch line. A deep swale can result in high maintenance costs if it becomes necessary to haul in fill to bring the trench back up to grade. If fill is not brought in, water running across the pipeline right-of-way can be intercepted by the pipe trench. Water running down the trench can cause extensive erosion of the backfill. This can possibly lead to areas of exposed pipe, or floating pipe. Either can be expensive to repair, particularly if access to the area is restricted. In extreme cases, floating pipe can require depressurizing the pipeline or it can result in a pipeline shutdown.

In permafrost terrain, particularly discontinuous permafrost, a deeper trench may have the benefit of reducing the pipeline's operational differential thaw settlement. Reduced thaw settlement may result because trenching through the uppermost 300 mm to 600 mm of mineral soil may remove high ice content soil that is often present just below the organic layer.

5.5.4.2   Imported Fill

The imported fill option is similar to the deeper ditch option in that local mineral soil is used to provide the negative buoyancy required to hold the pipe down. The difference is that instead of using the mineral soil excavated from the ditch bottom, mineral soil is imported from a nearby borrow source.

One advantage of using imported fill is that a deeper trench is not required. This saves on excavation costs, and also reduces the potential for post-construction ditch backfill settlement. The main disadvantage is the relatively high cost of opening up a separate borrow source, including permitting and access etc., along with haul costs to transport the fill from the borrow source to the pipeline. These costs generally exceed the cost of alternate types of buoyancy control measures.

5.5.4.3   Granular Bag Weights

This buoyancy control option involves the use of one of several different patented types of bags. The bags are typically made of some type of long lasting polypropylene or similar type material. Two types of bag weights are shown in Figures 8 and 9. The bags may have several different compartments, depending on the type and the size. The bags are produced at the particular company's manufacturing facility and shipped either empty or full to the job site. Unfilled bags are lightweight and thus inexpensive to ship. Empty bags are filled at a suitable borrow site(s) near the pipeline corridor. The fill material usually is a granular soil because of its higher unit weight. For field filled weights, such as the type shown in Figure 8, a hopper system (Figure 10) is used to fill the bags. Filled bags are placed over the pipeline. Some types are placed after the pipeline has been lowered into the ditch similar to concrete saddle weights. Others can be pre-attached to the pipeline before lowering in of the pipeline. This is advantageous where the trench is filling with water.

Additional bag weights can be shipped and filled in only a few days. This provides greater flexibility than concrete weights which would generally require more time to be poured, cured, and shipped from a distant manufacturing facility.

Figure 8: PipeSak™ Weight with 'Pinched' Base (from PipeSak™ Inc.'s website, downloaded May 19th, 2003.)

Figure 9: Bag Weights in Wet Ditch (courtesy Allen Brown of AHB Consulting Ltd.)

Figure 10: Hopper for Filling PipeSak™ Weight (from PipeSak Inc.'s website, downloaded May 19th, 2003.)

Bags are available for most pipe diameters off the shelf. Custom made bags also can be produced. Depending on the manufacturer, the bags may or may not be strapped to the pipeline. Thus, some may be installed prior to lowering-in, while some are installed following lowering-in.

The bags can be placed slightly further apart than equivalent mass concrete weights because the water saturates the granular fill, thus slightly reducing the displaced volume. Another advantage is that the weights have a lower center of gravity than concrete weights. This virtually eliminates the possibility of the weights "tipping" over, which has been a problem for top heavy concrete weights in some situations.

One point to consider when using bag weights is that they typically require a wider trench vs. concrete weighting for the larger pipe diameters. This may be partially offset by a reduced ditch depth vs. concrete weights. There may be a construction cost impact if the required ditch width exceeds the capability of the contractor's preferred trenching machines. However, in many areas where buoyancy control is utilized, backhoes are used to excavate the trench in order to achieve the required trench depth and width. In this case, a slightly wider trench for a granular bag weight has little cost impact.

Additional information on bag weights is available in the paper by G.W. Connors.

5.5.4.4   Geotextile Swamp Weights

Geotextile swamp weights (GSWs) are not in common use. This weight type was tested by a Canadian company in the 1990s and found to be suitable for smaller pipe diameters. They were used on several projects but none recently that the writer is aware of. GSWs like bag weights also use a geotextile fabric but GSWs are not pre-manufactured weights. The geotextile fabric, generally a non woven medium weight polypropylene is simply draped over the pipeline and across the entire pipeline trench. A mineral soil backfill is then placed over the fabric to a specified thickness. The fabric is then folded over the mineral soil so it encapsulates the soil at the top and bottom; the ends are left open. (See Figure 11) The GSW adds to the effective weight on the pipeline by making use of the mineral soil beside the pipe in addition to the soil over the pipe.

The mineral soil used to backfill the weights can be the material excavated from the bottom of the pipeline trench below the organic layer or it can be local fill from a nearby borrow source. Any type of higher density soil is suitable. The fill does not have to be granular; it can be clay. Silt backfill is not recommended due to its lower submerged density and the higher possibility of soil loss out the ends of the GSWs.

These weights have the advantage of low material and shipping costs. A disadvantage is they do have higher labour costs as it takes several people to lay out the geotextile, pin it, backfill it, and do the final folds. The labourers generally

186

have to enter the trench, which means suitable backsloping of the trench walls and adherence to other trench safety requirements are required.

GSWs will generally require a closer center to center spacing than concrete weights, but this is dependent on the height and type of mineral fill placed inside the weight.

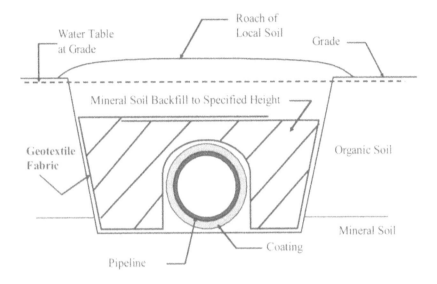

Figure 11: Geotextile Swamp Weight

5.5.4.5   Plate Weights

Plate weights (PWs), like GSWs, are not a common weight type and are not in common use. This weight type was tested by a Canadian company on two different pipe diameters. Plate weights are a thin, high-density plastic weight. They are made using high-pressure injection moulding and a pre-made stainless steel mould. The moulds are expensive to produce, but once made, thousands of plate weights can be produced from a single mould. A lengthy lead time, up to two years, may be required to produce the mould and the weights. Also, there are very few manufacturing facilities in North America with the capability to work with the large moulds that are needed as most high pressure injection facilities handle only small parts.

The finished PWs can be carried in the back of a pick-up, and placed on the pipe by two labourers. The plates are simply placed over the pipe, much like a wide-brimmed hat, and then backfilled with mineral soil. (See Figure 12) The mineral soil can be the mineral soil from the bottom of the trench or locally sourced imported fill. Similar to GSWs, the mineral soil adjacent to the pipe contributes to weighting the pipeline in addition to the mineral soil present above the pipe.

Similar to GSWs, PWs may require a tighter spacing than concrete weights, depending on the height and type of mineral fill placed on top of the weight.

For both GSWs and PWs, backfill stability is important to the ongoing effectiveness of the weight. A loss of ditch backfill will result in reduced factor of safety against pipe uplift due to buoyancy. Thus, water flow down ditchline and subsequent erosion of the trench backfill must be minimized. Regularly spaced berms and/or trench plugs can be used to reduce ditchline erosion by directing ditch water out of and away from the ditch.

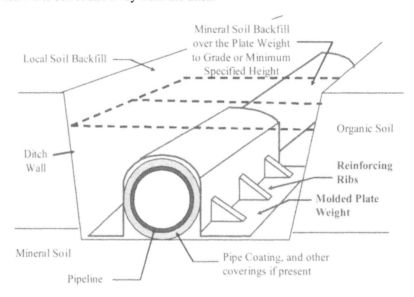

Figure 12: Plate Weight

188

5.5.5 Anchors

Anchors differ from the previous weighting methods in that they do not rely solely on mass to hold the pipe down. Anchors utilize the strength of the underlying mineral soil combined with the mass of the soil above the anchor to resist pipe uplift. Anchors basically consist of two metal rods screwed, drilled, or placed into the subsurface soil, one on either side of the pipeline. The metal rods extend up to approximately the mid-point of the pipeline. A saddle is then draped over the pipeline and attached to the rods on either side. The pull-out resistance of the anchors is what provides the resistance to the buoyant force of the pipeline. Figure 13 shows a typical screw anchor set-up.

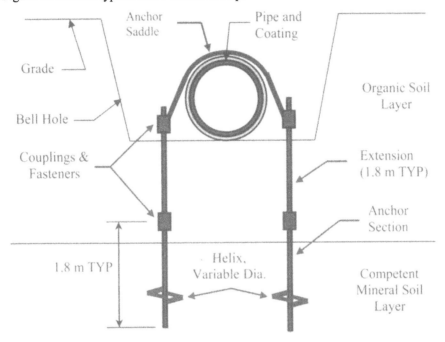

Figure 13: Schematic of Typical Screw Anchor Set-Up

Anchors can be installed in most soil conditions, but the anchor type will vary with the subsurface conditions. The difference in anchor types is the method of advancing or placing the steel rod, and how the rod is held in place. Anchors, like saddle weights or some soil weights, are difficult to place if there is excess water in the trench that cannot be pumped out. In these situations, river weights or concrete coating should be used.

The main advantages of anchors are:

1. An anchor set can carry very high loads. This allows for the spacing between anchor sets to be three to seven times greater versus concrete or bag weights.

2. Anchor shipping costs are less than concrete weighting because of their lower weight and volume. Because shipping costs are low, any unused anchors can be easily and inexpensively returned to the warehouse.

3. There is greater ease in shifting anchor locations because they do not need to be strung out along the RoW prior to trenching and lowering in. Anchor locations can be finalized following trenching, allowing for an inspection of the soil conditions.

Two main disadvantages of anchors are:

1. The proper anchor type and the required anchor length greatly depend on the soil conditions below the pipeline. The presence of coarse granular material, bedrock, or permafrost in the subsurface soils must be known. It is also desirable to know the depth to the stiff soil the anchor will be founded in to allow for pre-construction ordering of the correct quantities of extensions.

2. Anchors require specialized installation equipment and, depending on the anchor type, one to three qualified installation personnel. Depending on the number of anchors and their relative location along the pipeline, the specialized installation personnel may have a number of standby days.

There are three anchor types discussed below, screw anchors, grouted anchors, and freeze-back anchors. Only screw anchors are currently in typical use on overland pipelines.

5.5.5.1   Screw Anchors

Screw anchors for pipelines are used in the United States, Western Canada, and parts of Asia. As shown in   Figure 13, screw anchors consist of a leading steel rod with single or multiple helices welded onto the rod. Helices are available in a range of diameters from 100 mm to 400 mm. They are also available in single, double, or triple combinations. The load capacity of an anchor generally increases with the increasing helix surface area. As a general rule, each helix can provide a maximum ultimate load of 180 kN. Actual load capacity will depend on the anchor configuration and the installed torque, which is in turn dependent on the soil conditions. Load resistance is derived from a combination of the bearing capacity of the soil in contact with the helix and the shear strength of the soil acting on a cylindrical surface above the perimeter of the helix. The correct pitch of the helix is critical to proper installation and in obtaining the design anchor load. The anchor must advance into the soil without excessively disturbing the soil column above the helix. Excessive soil disturbance will result in reduced anchor capacity. Thus, anchors must be produced by approved suppliers.

A screw anchor load calculation example is shown in Section 5.14.3. Screw anchor suppliers also can provide design manuals and software programs for estimating anchor capacities based on soil strength. Robertson and Curle and Doering and Robertson describe screw anchors in some detail in their papers.

The lead steel rod and the extensions typically consist of 38 mm or 44 mm square shaft steel. The rods are generally 1.8 m or 2.1 m long. There are 0.9 m long extensions available; these are typically used to complete an installation where the minimum installation torque is close to the required value. For pipelines, specifications generally require a minimum embedment depth of two full length rod sections, one anchor lead section with the helix and one extension.

The anchors are installed by rotating the leading rod with the helix through the soft organic soils and into an underlying stiff mineral soil. The two leading rods can be installed singly or together, (see Figures 14 and 15) depending on the equipment type. Rod extensions are added as necessary until the specified minimum installation torque resistance is achieved (see Figure 14) over a minimum length of three helix diameters. The anchor installation is completed by placing a saddle over the pipeline and attaching the saddle to each anchor (see Figure 16).

Construction specifications should require that a certain number of anchors be pull tested, on the order of every 5th anchor see Figure 17. Pull loads should be a minimum of 115% of the design load. The pull test is usually done with a track hoe. A large hoe is required where anchor capacities are high. This can sometimes be a problem as the available hoe pull load is reduced if the boom must be extended a significant distance, as is the case for anchors on the far side of the trench.

Screw anchors cannot be used in areas of shallow bedrock or where the underlying soil consists of coarse granular materials that prevent helix penetration.

191

Figure 14: Dual Drive Head for Installing Screw Anchors

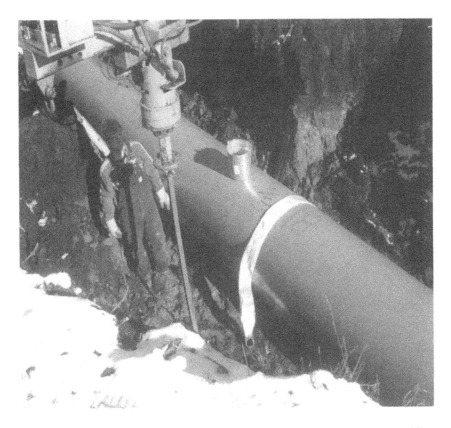

Figure 15: Single Drive Head for Installing Screw Anchors (courtesy Allen Brown of AHB Consulting Ltd.)

Figure 16: Installed Anchors with Anchor Saddle (courtesy Allen Brown of AHB Consulting Ltd.)

Figure 17: Anchor Pull Test

194

5.5.5.2 Grouted Anchors

An alternative to screw anchors in difficult ground conditions is the grouted anchor. They are commonly used in the building construction industry as tieback anchors for retaining walls and can hold large loads. However, these anchors have been used infrequently on pipelines. Grouted anchors were tested on a pipeline by a large Canadian pipeline company. The trial demonstrated that grouted anchors are technically feasible for use in pipeline construction. It also demonstrated that grouted anchors are capable of safely holding large uplift loads, as is routinely done by tieback anchors. This would potentially allow for a larger spacing between grouted anchors vs. screw anchors. However, as noted previously, the maximum available pull load may dictate the maximum anchor capacity.

Grouted anchors differ from screw anchors in that they do not use a lower helix, only straight rods. The steel rods are placed into undisturbed competent soil below the upper organic or soft soils. (See Figure 18). The rods can be placed into a pre-drilled hole, either cased or uncased or they can be driven directly into the ground. The rods are then grouted into place using a prescribed minimum volume of grout combined with observations on grout return. When placed into granular soils or bedrock, the anchors can provide excellent load carrying capacity.

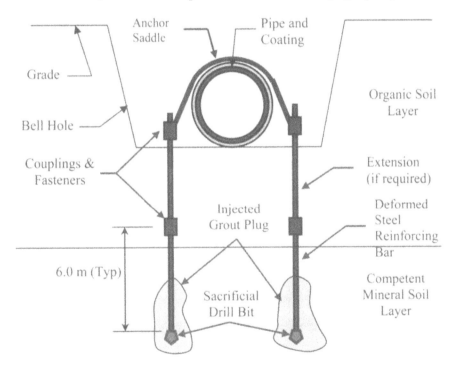

195

Figure 18: Schematic of Injected Grouted Anchor Set-Up

Figure 19: Anchor Installation Drill (From www.johnhenryrockdrill.com)

The type of installation equipment differs from screw anchors. While screw anchors are installed using either a backhoe equipped with single or dual drive heads or a 'picker truck with a single drive head, grouted anchors are installed using a compressed air powered drill rig. Usually a downhole hammer powered by compressed air is utilized. There are many commercial 'airtrack' drill rigs available, but the drill equipment also can be mounted on a backhoe. (See Figure 19)

The use of a downhole hammer allows grouted anchors to be installed in very hard ground. The hammers can penetrate through granular soils, cobbles, and bedrock.

Some items to consider when grouted anchors are being proposed for use on pipelines:

- Special expertise and equipment outside of typical pipeline expertise is required to install the grouted anchors. For example, the air powered drill might be mounted on a "John Henry" (where it can be boomed out over the pipeline trench).

- The productivity (number of anchors installed per day) of a grouted anchor installation crew likely will be less than that of a screw anchor crew.

- The grout requires some time to setup prior to testing whereas screw anchors can be pull tested immediately following installation.

5.5.5.3 Freeze-Back Anchors

Freeze-back anchors, or sometimes referred to as slurry anchors, are only suitable in areas where there is cold permafrost. They are similar to the freeze-back piles used in northern building construction. In building construction in permafrost ground, a steel pile is placed into a predrilled hole. The hole is then backfilled with a sand/water slurry. The frozen ground causes the slurry to freeze and strong adfreeze bonds are developed between the ground and slurry and the slurry and pile. The adfreeze strength depends in part on the type of slurry used and the final slurry temperature. Smith and Hernadi present the results of some tests on the uplift capacity of pipe piles using sand-water slurry. The paper presents results of both field and laboratory tests.

For freeze-back anchors, a deformed steel reinforcing bar could be used in place of the pile. The use of a steel bar vs. the pile would reduce steel costs as well as facilitate the anchor to saddle connection. (See Figure 20) The bar may have an end plate to increase the load bearing area. The adfreeze skin friction provides the uplift resistance against the pipeline's buoyant force. Freeze-back anchors thus are more similar to the grouted anchor than the screw anchors in that they do not require a helix. They differ from the grouted anchor, however, in that they do not rely mainly on the strength of the native soil.

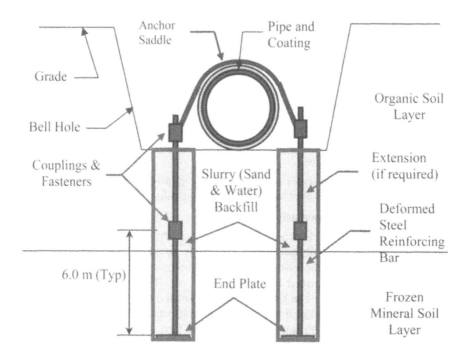

Figure 20: Schematic of Freeze-Back Anchor Set-Up

Freeze-back anchors have not been used on a northern pipeline to this writer's knowledge. However, they have been tested. The Canadian Arctic Gas Study tested slurry anchors in 1974 (EBA June 1974). The Mackenzie Gas Project (MGP) also tested freeze-back anchors in 2005. In the EBA test (there is no public data on the MGP tests) loads from 154 kN to 323 kN were carried by seven slurry anchors with effective lengths that varied from 3.4 m to 3.9 m. The corresponding ultimate skin friction varied from 94 kPa to 220 kPa. These capacities were developed in permafrost where the average ground temperature was minus 2 degrees Celsius (°C) to zero °C. The report stated that slurry anchors were a realistic buoyancy control alternative where it is assured the average ground temperature would not exceed minus 1.0 °C.

An important consideration for freeze-back foundations is the effect of long term creep. The EBA report presented an equation to be used for determining safe anchor capacities based on service life and maximum tolerable upward creep. Long term creep considerations will significantly reduce safe anchor loads from the ultimate load if creep is to be kept to less than a few millimetres.

Similar to grouted anchors, slurry anchors are installed using a compressed air powered drill rig. Usually a downhole hammer powered by compressed air is

198

utilized. There are many commercial 'airtrack' drill rigs available, but the drill equipment also can be mounted on a backhoe. (See Figure 19)

The use of a downhole hammer allows slurry anchors to be installed in very hard ground. The hammers can penetrate through granular soils, cobbles, bedrock, and the thick ground ice that may be present in some permafrost areas.

If a designer is planning on using freeze-back anchors, additional field testing may be necessary to address the issue of the maximum allowable anchor load based on soil type and soil temperature (colder ground temperatures will increase allowable anchor loads) and allowable load vs. ultimate based on acceptable creep rates.

Other issues to be considered for freeze-back anchor design include:

- Embedment depths required with due consideration to long term ground temperature changes resulting from a change in the ground thermal regime (effect of tree or bush clearing and compaction or loss of the organic cover) and from the pipeline operating temperature range(s) over the pipeline's design life

- Installation equipment type(s) and availability, personnel expertise, daily productivity and costs relative to other buoyancy control methods

- Anchor proof testing method and timing as related to freeze-back time (delayed anchor testing may affect pipeline construction methodology and efficiency)

- Need for long term anchor uplift (creep) or ground temperature monitoring

- Steel corrosion rates in permafrost

5.5.5.4 Anchor Corrosion

As anchors consist of steel parts, their long term corrosion must be considered when choosing anchor components. Galvanized steel is one option. A second is to assume an annual corrosion rate applicable to the local terrain and ensure there is sufficient additional steel in the rods and couplings to sustain the required load over the design life even after corrosion has occurred. The exception is the bolts used in the couplings and the top-of-anchor terminators. These are usually coated with xylan or another approved coating or are galvanized.

5.5.6 Summary of Various Buoyancy Control Applications

The local subsurface soil types can affect the preferred buoyancy control option. Table 2 presents a summary of the possible terrains where each buoyancy control type could be considered.

Table 2:   Potential Buoyancy Control Method for Various Field Situations

Buoyancy Control Method	Terrain Condition							
	Shallow Organics Over		Deep Organics Over				Wet Organic Areas or Minor Water Crossings	Major Water Crossings
	Unfrozen Soil or Warm Permafrost	Cold Frozen Ground	Unfrozen Soil or Warm Permafrost	Cold Frozen Ground	Shallow Bedrock	Granular Soils		
Deeper Ditch	XX	X						
Imported Fill	X	X						
Plate Weights	X	X						
Geotextile Swamp Weights	X	X						
Screw Anchors	X		XX					
Freeze-Back Anchors		X		XX	X^1			
Grouted Anchors			X		X	X		
Concrete Swamp Weights	X	X	X	X	XX	XX		
Concrete River Weights	X	X	X	X	X	X	XX	
Bag Weights	X	X	X	X	XX	X	X^2	
Concrete Coating			X	X	X	X	X	XX

XX – Preferred option, but dependent on local economics, local terrain, and lengths of buoyancy control required.

1 - if cold permafrost

2 - if pre-attached

Concrete Set-on Weights:

Advantages:

- can be installed in any type of soil terrain
- uses less 'outside' (non prime pipeline contractor) labour
- uses equipment the prime pipeline contractor already has on site
- familiar to all prime pipeline contractors
- can be placed rapidly
- available from a variety of manufacturers

Disadvantages:

- high transportation cost (from the manufacturer to the job site or the high cost of the remote batch plant) and increased traffic on the right-of-way
- require extensive lead time to make for winter projects
- require large storage areas
- reduced center to center spacing vs. anchors
- once purchased, often difficult to return for credit
- require a continuous wider and deeper ditch vs. anchors or imported fill
- require a partially dewatered ditch
- subject to overturning if not properly placed and there is a lack of adequate lateral support
- require heavy equipment to lift and place
- potential for pipe damage if dropped
- decreased worker safety (lifting of heavy loads)
- add a large amount of weight (additional stress) to the pipe if the water table drops
- require extensive 'concrete quality' granular material to build (may be an issue in some areas)

Continuous Concrete Coating:

Advantages:

- can be placed in any type of soil terrain
- can be placed in flooded ditches
- uses installation equipment the prime pipeline contractor already has on site
- can be dragged into place (wide watercourse crossings)

- can be floated into place
- provides additional protection to the pipe and coating
- is not subject to overturning
- continue to provide buoyancy control even if all the backfill over the pipeline is lost, usually due to erosion

Disadvantages:

- require on-site hoarding and heating for remote winter projects
- require a continuous wider ditch vs. anchors
- often require 'outside' (non prime pipeline contractor) to prepare forms and place the concrete
- requires equipment capable of lifting combined mass of pipe and concrete (unless floated into place)
- disrupts the lowering-in procedure
- require extensive 'concrete quality' granular material to build (may be an issue in some areas)

Concrete Bolt-on Weights:

Advantages:

- can be placed in any type of soil terrain
- can be placed in flooded ditches
- uses less 'outside' (non prime pipeline contractor) labour
- uses equipment the prime pipeline contractor already has on site
- are not subject to overturning
- familiar to all contractors
- continue to provide buoyancy control even if all the backfill over the pipeline is lost, usually due to erosion

Disadvantages:

- reduced center to center spacing vs. anchors
- high transportation cost (from the manufacturer to the job site or the high cost of the remote batch plant) and increased traffic on the right-of-way
- require extensive lead time to make for winter projects
- reduced flexibility in adjusting weight locations due to the transportation costs and logistics
- require large storage areas
- once purchased, often difficult to return for credit

- require spacers (wood lagging commonly) to maintain desired weight spacing during installation
- require a continuous wider and deeper ditch vs. anchors
- requires equipment capable of lifting combined mass of pipe and weights
- disrupts the lowering-in procedure
- relatively slow to install
- require extensive 'concrete quality' granular material to build (may be an issue in some areas)

Deep Ditch:

Advantages:
- low cost
- no specialty equipment or labour
- can be implemented quickly in hoe excavated ditch locations
- may reduce post construction pipeline thaw settlement (where high ice soils are concentrated near regular ditch bottom)

Disadvantages:
- only suitable for areas where mineral soil is present within some portion of the base of the ditch
- may be increased post construction settlement and erosion of the ditch backfill

Placement of Select Backfill, Geotextile Weights and Plate Weights:

Advantages:
- may be cost effective if able to use local material and hauling costs are minimal
- does not require extensive lead time (except plate weights)
- does not require a wider or deeper ditch
- can double as pipeline padding

Disadvantages:
- only suitable for areas where mineral soil is present within some portion of the base of the ditch
- more labour intensive
- may require development of borrow sites and associated environmental impact and permitting requirements
- requires large amounts of hauled fill

- require additional equipment such as earth movers or dump trucks to move the fill
- may require a partially dewatered ditch to obtain a suitable in-place soil density
- lateral pipe movement may result in slippage of soil off the pipeline
- may require hauling away of ditch spoil

Gravel Filled Bags:

Advantages:

- can be installed in any type of soil
- mainly uses equipment contractor already has on site
- able to use local material if available and thus reduce hauling costs
- if included in the contract, able to return excess bags
- does not require extensive lead time to fill/add additional bags, thus highly flexible
- are less subject to overturning vs. concrete set-on weights
- low transportation cost to move empty bags between construction spreads

Disadvantages:

- reduced center to center spacing vs. anchors
- require extensive granular material of the proper gradation to fill the bags
- require additional equipment to fill bags (or higher hauling costs if pre-filled)
- require a continuous wider ditch vs. anchors or concrete weights
- may require a partially dewatered ditch
- potential for pipe damage if dropped
- decreased worker safety (lifting of heavy loads)
- add weight (additional stress) to the pipe if the water table drops

Screw Anchors:

Advantages:

- economical due to reduced transportation cost and increased spacing between anchors relative to other methods
- economical to move between construction spreads
- ability to obtain additional stock, depending on quantity, in a relatively short time frame

- ability to return excess (non-custom) stock, usually with a nominal re-stocking charge
- does not add weight (additional stress) to the pipe
- reduced ditch dimension between anchors and thus higher ability to excavate the majority of the trench using chain or wheel ditcher instead of backhoes (increased ditching productivity)
- can be installed either immediately after lowering-in or after partial backfilling of ditch (leave anchor locations open)
- reduced congestion on the row from less truck traffic (hauling trucks) and less required storage space

Limitation:

- limited, based on current design procedures, to a maximum load per anchor of 75 kN, or 150 kN per assembly

Disadvantages:

- cannot be installed in shallow bedrock, coarse gravel/cobble areas, or hard (cold) permafrost
- additional expertise (anchor technician) required to install (may be additional downtime for the specialty labour)
- additional equipment required (i.e. drive heads and modified hoe)
- more 'outside' (exposed to the weather) labour required
- requires a dewatered ditch unless special procedures and equipment are employed
- requires a safe bell hole to allow worker ingress
- not familiar to all contractors

Grouted Anchors – similar to screw anchors except:

Advantage:

- increased maximum load per anchor to the point where pipe displacement or pipe strain would govern the maximum spacing

Disadvantages:

- additional expertise (i.e. grout technician and drillers) required to install (may be additional downtime for the specialty labour)
- additional equipment required (i.e. specialty drill, grouting tanks, hoses etc)
- specialty materials required (anchor rod and connections)
- delay in load testing anchors until grout has sufficiently cured

- reduced productivity, which may lead to the requirement for an additional crew

Freeze-back Anchors – similar to screw anchors except:

Advantage:

- can be used in cold permafrost

Disadvantages:

- additional expertise (i.e. slurry technician and drillers) required to install (may be additional downtime for the specialty labour)
- additional equipment required (i.e. specialty drill, tanks, hoses etc)
- specialty materials required (anchor rod and connections)
- delay in load testing anchors until slurry has sufficiently frozen
- reduced productivity, which may lead to the requirement for an additional crew

5.6 Buoyancy Design Forces

The various forces affecting pipeline buoyancy are depicted in Figure 21. Forces acting upwards include the pipe buoyancy, pipe coatings that may be buoyant in water such as extruded polyethylene or insulation, or pipe protection such as wood lagging or foam. Forces acting downward include such items as the pipe itself if it is made of steel or another material heavier than water, the product in the pipe if they are heavier liquids such as oil or bitumen, pipe coatings heavier than water such as fusion bond epoxy or a three layer coating, pipe protection such as rockguard, and the weight and strength of the soil over the pipe.

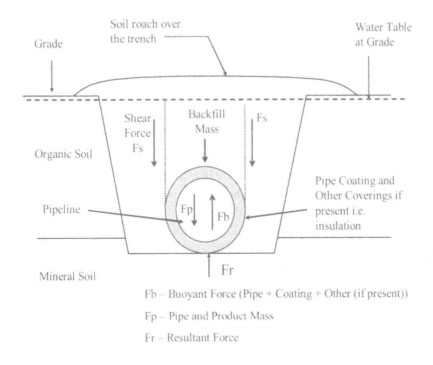

Fb – Buoyant Force (Pipe + Coating + Other (if present))

Fp – Pipe and Product Mass

Fr – Resultant Force

Figure 21: Forces Affecting Pipeline Buoyancy

Further notes on the forces are:
1. Forces acting upwards:
   - pipe buoyant force - based on Archimedes Principle = displaced volume times the density of the displaced liquid.
   - wood lagging (between river weights) = volume lagging times submerged mass
   - extruded polyethylene coating and other coatings = volume coating times submerged mass
2. Forces acting downwards:
   - pipe (if pipe material is heavier than the displaced liquid (typically water)) = volume pipe wall times mass
   - product in the pipe = volume pipe product times mass
   - soil mass above the pipe = volume of soil times submerged mass
   - strength of soil mass above the pipe = shear strength along the vertical slip surface proportional to a defined overburden stress and the shear friction angle
   - concrete coating or concrete weight = volume concrete times submerged mass
   - pipe coating or pipe protection = volume coating times submerged mass (when heavier than the displaced liquid)
3. Neutral forces: some of the above items may not be included in the buoyancy calculations at all times.
   - organic layer mass – If the only soils above the pipe consist of pure or near pure organic soil, the weight of any soil above the pipe may be neglected. The designer may still however include a frictional force from the organic soil layer due to its fibrous nature and "interlocking" of the root mass.
   - mineral soil – The mass of the mineral soil above the pipe would be excluded at a wet watercourse crossing, or in very wet swampy areas as it is necessary to sink the pipe to the ditch bottom during construction, prior to any backfilling. Thus, a sufficient amount of buoyancy control needs to be added without consideration of any soil weight.
   - dry gas product – Dry gas in pipelines is usually considered to be weightless. The designer however, may at his discretion give consideration to the mass of entrained liquids or other products in the gas stream as these can at times be significant. The designer must consider though the timing of product flow vs. when the pipe may be subject to buoyancy forces. Buoyancy calculations must

consider the case where the pipelines have not yet been placed into operation and are empty, but go through a period where a high water table may occur. Examples are the first post-construction thaw season following winter pipeline construction in northern areas, or heavy rainfall or flooding events that occur shortly after construction.

- liquid pipeline product –Although the product in liquid carrying pipelines or wet gas pipelines does have weight, as per the previous bullet, buoyancy calculations must consider the cases where the pipelines have not been placed into operation prior to being subject to buoyancy forces.

- soil roach – The roach of excess backfill soil placed over the pipeline, (extra soil due to the soil volume displaced by the pipe plus bulking of the soil as it is excavated) whether it be mainly organic soil or an organic/mineral soil mixture, is neglected in the buoyancy calculations. The roach is, however, assumed to partially compensate for post construction settlement of the loose backfill. The mass of the soil roach cannot always be counted on as it may not always be properly centered over the pipeline trench, or it may be eroded by surface runoff following rain events.

5.6.1 Buoyancy Calculation Input Values

Natural mineral soil and organic soil density values range widely. Friction angles also vary widely. There are even wider variations to consider when noting that the values used must be representative of the highly disturbed condition of the backfill replaced into an excavated pipeline trench. The actual density value used is subject to local conditions, judgement, and degree of conservatism in all the buoyancy factors, plus the level of risk willing to be accepted. The benefit of using higher soil densities or friction angles must be assessed against the risk of pipe exposure. The added cost of a little more buoyancy control is usually much less than post-construction costs to remediate a pipe exposure.

5.6.1.1 Mineral Soil Density

Soil density values range widely. Soil wet density can range from in the order of 1600 to 2200 kg/m^3. Ideally, the actual unit weight for the site specific replaced trench material would be used in buoyancy calculations. However, site specific values are difficult to accurately determine. It is thus necessary to adopt a number for design purposes. In order to avoid having to field verify design values, a conservative density needs to be adopted.

A mineral soil density value in the order of 1650 kg/m^3 is a value accepted by some in industry. Simmonds and Thomas used a density of 1650 kg/m^3, while Couperthwaite and Marshall quote a value of 1600 kg/m^3. A value of 1650 kg/m^3

could be considered to be conservative, based on typical undisturbed bulk soil densities. However, as stated previously, it must be realized that the backfill soil is highly disturbed for at least the initial two to three years after construction. Also, in the situation where there is an organic layer 300 mm thick or more, a lower density value provides an allowance for the loss of a portion of the mineral soil component that ends up directly overlying the pipeline. This may occur where the organic layer only partial mixes with the lower mineral soil during trench excavation and backfill with the result a higher proportion of mineral soil is placed beside the pipeline vs. above it. A low density value also provides, for some soil types, an allowance for some loss of the roach over the pipeline, or an improperly centered roach.

The chosen density of mineral soil has two main effects. One, it will affect the thickness of mineral soil required above the pipe before buoyancy control is needed. Two, the density of mineral soil will have a direct impact on the spacing of soil type weights, such as Geotextile Swamp Weights or Plate Weights where the insitu soil is used.

The mineral soil density does not usually affect the spacing of concrete weighting, soil anchors, or granular filled weights. The usual conservative assumption is that there will not be any mineral soil present in the ditch backfill where these weight types are used. The designer may at his discretion, however, choose to include any mineral soil mass in the buoyancy calculations.

5.6.1.2   Organic Soil Density

The density of organic soil, like mineral soil, can be expected to vary considerably from location to location. A significant variable affecting organic soil density is the organic/mineral content proportions. Simmonds and Thomas quote a density of 1121 kg/m^3 for a soil with only 36% organic content while Couperthwaite and Marshall present values of 1450 kg/m^3 to 1580 kg/m^3 for backfill with organic contents ranging from 100% to 50%. These results indicate a large possible range in the final disturbed density of the organic soil. Again, a conservative value is recommended for use in design as the actual insitu density is usually not measured or known. This is considered to be a reasonable approach for long linear pipelines where the organic soil types likely will vary widely and proper density testing of the disturbed organic layer is difficult. A value of 1050 kg/m^3 is recommended. This value is based on the assumption that there is little to no mineral soil content within the organic soil.

5.6.1.3   Backfill Shear Resistance

The internal friction of soil or organics also provides additional resistance to pipe buoyancy. Undisturbed shear resistance values for various soils are available in many soil text books. However, as the replaced trench soil is highly disturbed and may contain numerous voids, lower shear values should be used in design. Another alternative is to ignore shear entirely for buoyancy calculations in

210

overland areas. This would be a very conservative approach particularly for small diameter pipelines.

Simmonds and Thomas provide a summary of laboratory test results for various mixtures of organic and mineral soils. Back calculated friction angles ranged from 29 degrees for a silty sand to as high as 62 degrees for a high organic content soil. Based on these test results as well as many years of buoyancy control design success incorporating a shear component, it is recommended that some shear resistance be included in overland buoyancy control calculations. Again, a conservative value in the order of 10 to 15 degrees is recommended. This value can be applied to both the disturbed mineral soil and the upper organic soil. Higher values could be used, but as noted in the introduction to this section, the benefit of using higher friction angles must be assessed against the other design values, the chosen safety factor, and risk levels of pipe exposure.

It should be noted that as a percentage of the overall reduction of positive buoyancy, the shear angle has a minor effect on larger pipe diameters (> NPS 20) whereas the effect is more pronounced, about 22% increase in swamp weight spacing, for smaller pipelines. This is demonstrated in Figure 22.

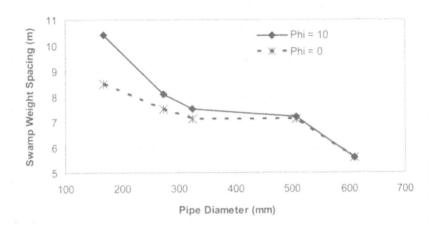

Figure 22: Effect of Backfill Shear Strength on Pipe Buoyancy

5.6.1.4   Density Values for Water and Other Items

The densities of ditch water and other components are also required to calculate pipe buoyancy. Suggested values for these items are:

- Use a density of fresh water of 1000 kg/m^3 at watercourses. If the water is brackish, a density proportionate to the salt content should be used.

211

- Use a density of water in muskeg areas of 1040 kg/m^3 with the water table at the ground surface. The higher density vs. fresh water allows for some suspended sediment in the water. Again, if the water is brackish, an increased density proportionate to the salt content above the 1040 kg/m^3 should be used.

- Use a density of 2250 kg/m^3 for concrete in swamp or river weights and in continuous concrete coating.

- Use a total density of 430 kg/m^3 for wood lagging placed around the pipe (often used between river weights)

- The density of the pipe wall material will depend on the type of pipeline. For a steel pipeline, the usual density is 7850 kg/m^3.

- For steel pipeline coatings, the supplier's actual provided density should be used. A total density of 950 kg/m^3 can be used for polyethylene pipe coating and 1440 kg/m^3 can be used for a fusion bond epoxy coating. (The thickness of fusion bond epoxy coating is only a few millimetres and can be excluded from buoyancy calculations with little effect on the final values.)

- The density or densities of the product being shipped in the pipeline may also need to be factored into overland buoyancy calculations. However, the timing of when the pipe is put into operation needs to be considered in terms of what is the worst case for buoyancy control. i.e., will the pipe be subject to maximum buoyancy forces prior to the pipe being put into operation, or during planned or unplanned outages.

Some designers may choose to use a higher value for the density of water in muskeg areas. This is in effect applying a safety factor to one component. Similarly, a higher value can be used for concrete, depending on the quality control procedures in place. This is an individual decision, whether to apply safety factors to individual parts of the input parameters, or at the end to the final calculated pipe buoyancy.

5.6.2   Safety Factor

The safety factors that are typically used in buoyancy control are small. For pipelines crossing watercourses, the calculated negative buoyancy is increased by 10-20 % (i.e. the resultant downward acting forces on the pipe are 10-20 % greater than the upward forces). At watercourses where there is significant water velocity, the safety factor may be increased to facilitate lowering-in of the pipeline. Contractors may also add water to the pipe prior to installation to add weight. This additional weight does aid in maintaining the pipe at the ditch bottom during the backfilling procedure when caving of the trench walls and dropping of the trench backfill can sometimes cause the pipe to lift off the trench bottom, particularly in wide river crossings.

For overland buoyancy control, the calculated negative buoyancy is typically increased by only 5 %. These relatively low values take into account the conservatism in the chosen density values for the input parameters.

5.6.3 Soil Liquefaction

A unique situation can occur where soils are subject to liquefaction during seismic events. The soil no longer acts as a downward force on the pipeline, but instead acts as the liquid providing the buoyant force to the pipeline. In these cases, if it is desired to maintain pipe burial, a density that represents the density of the liquefied soil needs to be used in the buoyancy calculations. This will greatly increase the buoyant force relative to that provided by water only. For this scenario, the mass of the product in operating pipeline should be added into the calculations.

5.6.4 Pipe Stress

The effect of buoyancy control measures on pipe wall stress must also be considered. The cumulative effect of all external and internal forces on the pipe must be considered and must be less than or equal to the chosen limiting levels. Pipe stress can be a limiting factor on anchor spacing, whereas buoyant forces are usually the limiting factor for mass type buoyancy control measures. The anchor spacing calculations included in this chapter do take into account pipe stress, additional detailed stress analyses will be required for each specific project.

5.6.5 Operational Temperature

5.6.5.1 Warm Pipelines

An additional upward force, sometimes wrongly attributed to buoyancy, but one that does need to be considered in conjunction with buoyancy, is the force that results from pipe thermal expansion. The thermal expansion force may manifest itself at overbends and sidebends in the pipe. The expansion force at an overbend may be sufficient to lift the pipe above the surrounding grade. Thermal expansion can be especially problematic at bends in muskeg areas.

Similar measures used to control buoyancy may be placed at overbends to counteract thermal expansion forces. These measures may complement other buoyancy control measures used in that particular location. Thermal forces are not considered further in this chapter, but must be integrated into the overall pipe uplift design during detailed engineering on each particular pipeline project.

5.6.5.2 Cold Pipelines

A pipeline located in permafrost, such as many areas of Alaska or northern Canada, may be operated cold. Within the cold flow sections, the gas may be chilled to just below freezing before being injected into the pipeline. The natural loss of pressure and the Joules Thompson effect as the gas travels down the pipeline will chill the gas even further. Models are available which will predict the operating temperature of the pipeline along its length.

The actual operating temperature will affect decisions with respect to buoyancy control. One issue is: does a pipeline operated continuously below freezing require buoyancy control? Will a cold flow pipeline require a minimum of buoyancy control because the frozen soil around the pipeline locks it into the ground and thus withstands any potential positive buoyant force? This decision could translate into several million dollars of capital expenditure on buoyancy control within the cold flow sections if the frozen ground is considered to lock the pipe in, and thus negate the need for buoyancy control. There are three situations, however, that still have to be considered:

1. The construction of a long pipeline could take place over multiple winter seasons. Thus, there may be one or more summers during which the pipeline has been installed, but is not in operation.

2. Is the pipe operated below zero year round, or does the pipe operate warmer in the summer (but still maintaining an annual below 0 °C average.) A pipeline that runs warmer in the summer may allow the active layer to reach the pipe crown while also allowing a small thawed zone to develop around the pipeline. If both of these situations occur, the pipeline may no longer be "locked" into the ground. The pipeline may in this situation be subject to a sufficient buoyant force to overcome the downward forces and cause the pipe to rise to the surface.

3. There may be a time when the pipeline is not operating due to a maintenance issue or a failure, or following decommissioning. The time of year and length of the 'upset', and the initial ground temperature will determine if there is a possibility the ground around/above the pipeline will thaw. This can result in the pipeline floating in areas of wet, poor ground, if buoyancy control has not been installed. In a decommissioning situation, the decommissioning plan will need to consider any potential issues if the pipe does rise to the surface.

5.7 Buoyancy Control Spacing & Locations

This section presents examples of the spacing calculations for determining center-to-center (c. to c.) distances between the various weight types. This section expands on the basic philosophy presented in previous sections.

5.7.1 Concrete Weighting Spacing

The specific equations for calculating weight spacing are: (see also Simmonds and Thomas.

For a unit length of pipe, buoyant force is calculated in equations 1 and 2.

(EQN 1) $\quad \pi/4 * D_o{}^2 * \delta_{fluid} * g = F_{buoy}$

Where: $\quad \pi$ = a constant (3.14)

D_o = pipe and coating outside diameter

δ_{fluid} = fluid density

g = acceleration (gravity)

F_{buoy} = buoyant force per metre

Downward Force of the Pipe:

The downward force of the pipe is the combined weight of the pipe and coating. It should be noted that, some pipe coatings may be buoyant, as their density is less than that of water. The equation per unit length is:

(EQN 2) $\quad \pi/4 * (D_{os}{}^2 - D_{is}{}^2) * \delta_s * g + \pi/4 * (D_{oc}{}^2 - D_{ic}{}^2) * \delta_c * g = F_p$

Where: $\quad D_{os}$ = outside diameter pipe

D_{is} = outside diameter pipe

δ_s = pipe density

D_{oc} = Outside diameter coating

D_{ic} = inside diameter coating

δ_s = coating density

g = acceleration (gravity)

F_p = downward force of pipe and coating per metre

Net Downward Force from Shear Stress:

(EQN 3) $\quad 2 * \tau * D_p = F_{shear}$

Where: $\quad \tau$ = stress along each vertical slip surface ($K_o * \sigma * \tan\Phi$)

K_o = lateral earth pressure coefficient (0.5)

σ = overburden stress

Φ = internal friction angle

D_p = depth to the pipe spring line from grade

F_{shear} = downward force of shear stress per metre

Net Downward Force of the Weight:

(EQN 4) $\quad Vw * (\delta_w - \delta_{fluid}) * g = F_{wt}$

Where: $\quad V_w$ = volume of concrete weight

γ_w = concrete density

γ_{fluid} = fluid density

g = acceleration (gravity)

F_{wt} = net downward force of the weight

215

Weight Spacing:

(EQN 4) $F_{wt} / (F_{buoy} - F_p - F_{shear}) * FS$

Where: FS = Factor of Safety

<u>Example:</u>

A NPS12 steel uncoated pipeline with a wall thickness of 5.0 mm is placed in an organic area at 1000 mm below grade. Concrete weights with a density of 2250 kg/m^3 and a volume of 0.31 m^3 are used to weight the pipeline. Assume the density of the muskeg is 1050 kg/m^3, the internal friction angle is 15 degrees, and the water fluid density if 1040 kg/m^3. The necessary weight spacing is calculated to be:

$$F_{buoy} = \pi/4 * 0.323^{\,2} * 1040 * 9.81$$
$$= 835 \text{ kN/m}$$

$$F_p = \pi/4 * (0.323^{\,2} - 0.313^{\,2}) * 7850 * 9.81$$
$$= 384.4 \text{ kN/m}$$

$$\sigma = \tfrac{1}{2} * (1050 - 1000) * (1 + 0.323/2)^2 * 9.81$$
$$= 23.9 \text{ kN/m}$$

$$F_{shear} = 2 * \tau * D_p = 2 * 0.5 * 23.9 * \tan 15 * 1.1615$$
$$= 7.4 \text{ kN/m}$$

$$F_{wt} = 0.31 * (2250 - 1040) * 9.81$$
$$= 3679 \text{ kN}$$

$$F_p + F_{shear} = 384.4 + 23.9 = 408.3$$
$$F_{net} = (835 - 408.3) * FS = 426.7 * 1.05 = 448$$
$$\text{Weight spacing} = 3679 / 448 = 8.2 \text{ m}$$

5.7.2   Soil Weighting

The first soil weighting option discussed in Section 5.5.34 was extra depth trench. The amount of mineral soil required to offset the buoyancy of a particular pipeline can be calculated using equations 1 to 3 and adding in the weight of mineral soil above the pipe. The required height of mineral soil can be simply calculated by equation 5.

(EQN 5) $\{FS * F_{buoy} - F_p - F_{shear}\} / \{\delta_s - \delta_{fluid}\} * g = C$

Where: γ_s = soil wet density
           C = mineral soil thickness

As the shear force is dependent on the pipe burial depth, an iterative solution is required. Alternatively, for larger diameter pipes as noted previously, the shear

216

force can be ignored and this results in only a small effect on the overall downward forces.

The spacing for the soil filled bag weights or geotextile weights also can be calculated using equations 1 to 4. Bag weight suppliers can provide the submerged mass of their weights. The designer needs to set the volume and density of the soil to be used for geotextile weights or plate weights. Some suppliers will provide the required weight spacing if the needed input parameters are supplied.

5.7.3 Anchor Weighting Spacing

Anchor spacing calculations are calculated differently than for mass type buoyancy control. Soil density and shear values are not used in calculating anchor spacing. The three main criteria are limiting pipe wall stress, limiting anchor load to a typical maximum value for a specific diameter helix and limiting pipe uplift between anchors. The factors to be considered in the calculations include:

- longitudinal bending stress
- longitudinal membrane stress
- combined hoop and longitudinal stress
- combined stresses for restrained spans
- longitudinal stress due to sustained loading
- contact stresses at the anchor hold down strap on the pipeline

The calculations are based on linear-elastic stress rather than strain limits. The maximum stress due to pressure, buoyancy, thermal expansion and local support contact stresses is limited to the specified yield strength (SMYS) of the pipe material, or SMYS times a design factor. Stresses are calculated for both cases of restrained and unrestrained pipeline and the worst condition is assumed. A straight pipeline without overbends, sagbends, or sidebends is assumed. Only buoyancy loading is considered (no differential settlement loadings etc.).

Items used in the calculations are:

- a design factor appropriate to the project (generally 0.8 to 1.0)
- a maximum temperature differential appropriate to the project
- a zero negative buoyancy to determine anchor working loads
- a set maximum allowable uplift between anchors (100 mm has been used for many projects in Canada)
- the negative buoyancy contribution from any soil mass, any nominal muskeg mass, or backfill shear strength is often ignored in the calculations

217

Doering and Robertson discuss pipe stress in their paper. An example of the stress calculations for a NPS 24 pipe is given below. The limiting factor for anchor spacing for pipelines smaller than NPS 30 generally is pipe stress or the maximum 100 mm of allowable uplift between anchors. There is no strict technical basis for this uplift limit. It is set at 100 mm in order to keep pipe uplift below what is judged to be a reasonable amount. This limit could be adjusted for specific projects. Increasing the allowable uplift would increase the allowable anchor spacing in some cases. This benefit would have to be balanced against the reduced pipe cover between anchors, the higher pipe wall stress, and the higher anchor loads. It also should be remembered that for pipelines with a high temperature differential, the slightly uplifted sections may lift further as the pipe temperature increases.

For pipes larger than NPS 30, single helix anchor capacity is often the limiting factor vs. pipe uplift or stress limits.

Example: **Anchor load for a buoyant (i.e. gas) pipeline can be calculated using the following:**

Given: pipe OD = NPS 24, or 609.6 mm
pipe wall thickness (t) = 10 mm
pipe grade (SMYS) = 389 MPa
design pressure (P) = 9930 kPa
temperature differential = 50 °C
design factor (F) = 0.8
modulus of steel (E_c) = 207000 MPa
Poisson's ratio (v) = 0.3
coefficient of thermal expansion (α)= 1.2 E-5 per °C
pipe section modulus (Z) = 2778117 mm^3

Using previous equations, the pipe buoyancy is 2.98 kN/m, the dead weight is 1.45 kN/m with a resultant force F_r of 1.53 kN/m.

Hoop stress, $S_h = (P*D)/2t = 303$ MPa

Longitudinal bending stress, $S_b = (F_r * L^2) / 12Z = 51$ MPa

The longitudinal membrane stresses are calculated using the various cases of maximum and minimum temperature changes for both the pressured and unpressured pipe.

Longitudinal membrane stress, $S_L = v* S_h - E_c * \alpha * \Delta T$

Using the above, the maximum and minimum S_L for the pressured case is 90 MPa and -33 MPa respectively. For the unpressured case, the maximum and minimum S_L is 0 MPa and -124 MPa.

For a pipe supported in flexible slings, the maximum local stresses occur at the points of tangency and act in both circumferential and longitudinal directions. The screw anchor saddle is assumed to span 180 degrees and sits on top of the pipe. Therefore, the maximum stress occurs at the 3:00 and 9:00 o'clock positions.

Local contact stresses are calculated according to Roark and are:

$S_{contact,local} = 0.0092 *(F_r * La) / t^2 * \ln ((D/2) / t)$

where La = the assumed anchor spacing.

Longitudinal and circumferential stresses are combined using Tresca's criterion at the 12, 3, 6 and 9 o'clock positions. The maximum Tresca stress must be less than SMYS.

For this case, the maximum combined stresses occurs at the 12 o'clock position for the pressured case ($303(S_h)$ - -51 (for S_b) - -33 (for S_L) = 387 MPa for an La = 33.5 m. The 387 is just less than SMYS of 389 MPa.

Other requirements that must be met under Canadian code Z662 is that S_h less minimum S_L (303-(-33) = 336 should be less than 0.9*SMYS = 350 MPa, so it passes,

and $0.5*S_h$ + maximum of either S_b or $S_{contact,local}$ ($0.5*303 + 51 = 203$) should be less than SMYS * design, location, and temperature factors, or 311 MPa in this scenario. This requirement is also met.

The maximum assembly (two anchors) force is thus F_r * La = 33.5* 1.53 = 51 kN.

The maximum uplift between anchors is calculated to be 29 mm ($(F_r * La^4) / 384 EI$). This is generally an acceptable amount of upward movement.

Each anchor would have a working load of half the assembly force or 51/2 = 26 kN. Using a minimum safety factor of 2.5, the design load per anchor is 64 kN. The screw anchor components are then configured so that they are capable of holding the design load of 64 kN in this case.

Anchor capacity depends on the competency of the soil it is embedded into in conjunction with the size and number of the anchor helices. Single anchors with a nominal 200 mm diameter helix can have capacities ranging from 30 kN to 66 kN for loose to dense soils respectively (AB Chance Encyclopaedia). In the example above, a nominal 200 mm diameter single helix in dense soils, or a nominal 300 mm diameter helix in very stiff soils could carry the load (AB Chance Encyclopaedia). Anchor capacities can double with a doubling in helix diameter. Similarly, multiple helix anchors can again double or triple the capacities of single helix anchors depending on the number and size of the helices (AB Chance Encyclopaedia).

Past Canadian practice has been to use mainly single helix anchors. The advantage of this is that contact between the upper helices and the pipeline wall is reduced during installation. This reduces damage to the pipe and coating and

219

subsequent need for repairs. Typical helix sizes have been 200 mm, 250 mm, 300 mm and 400 mm. An anchor set with a single 400 mm helix on each lead section can typically maintain a working load of 150 kN if set into competent soil that provides the minimum required torque resistance during installation. Multiple helix anchors can certainly be used though. They may necessitate a more cautious installation approach while implementing appropriate measures to protect the pipeline.

The actual working load applicable to various soil conditions can be predetermined if the in situ soil properties are sufficiently known. Detailed descriptions of how to calculate anchor capacity can be found in Mitsch and Clemence and also Mooney et al as well as a variety of other sources. Reputable anchor suppliers can also provide assistance for project specific queries. The typical pipeline designer however, seldom has detailed test results on the underlying soil along many sections of a pipeline. Thus, a minimum anchor capacity is usually achieved by specifying a minimum installation torque rather than attempting to conduct detailed assessments of the subsurface conditions. (See also Section 5.15)

Once an anchor working load is calculated based on the specific pipe properties, the anchor size and anchor installation requirements can be determined. It is recommended that a minimum factor of safety of 2.5 be applied to the working load to get the design load. The design load is used to determine the materials strength for the anchor components. An anchor supplier can provide the maximum load capacity for their 'off the shelf' projects. The anchor type that meets the load requirements can be selected. Project specific anchors can be custom made, but this would be warranted for only large orders.

Anchor installation requirements are determined by applying a one-tenth ratio of minimum installation torque in foot-pounds (in Imperial units) versus the design load in pounds (Doering and Robertson). This ratio was derived from numerous field installations. In the example above, the minimum installation torque is 1/10 of the design load, 64 kN expressed in pounds (or 14340 * 0.1 = 1434 ft-lbs) or 1950 Newton-metres of torque. It is critical when installing anchors to a specified torque to ensure that the torque gauge is properly calibrated at the start of the project, and periodically through the installation phase. A high reading torque gauge will result in low capacity anchors. A low reading torque gauge could result in an excessive number of anchor extensions being required, or in excessive damage to the anchors. Some anchor suppliers can provide calibrated torque gauges to check the installers torque gauge. It is also important to remember that the calibration of a torque gauge may change when it is moved from one machine (hoe) to another. When torque gauges are switched between machines, the calibration should be checked.

It is recommended that a certain percentage of anchors be pull tested in order to verify the load capacity of the anchors, the soil, and the calibration of the torque

gauge. Some current specifications require that every fifth anchor in a section, or a minimum of one anchor for short sections, be pull tested. A minimum pull test load of 115 % of working load is recommended. The individual anchor pull test load in the example above would be
1.15*24 = 29 kN.

5.7.4   Spacing Comparison

It has been noted previously that anchors are capable of holding large loads. This load capacity allows for a relatively large center to center spacing vs. concrete weights or similar sized bag weights. Figure 23 provides a comparison of the spacing of concrete weights to screw anchors.

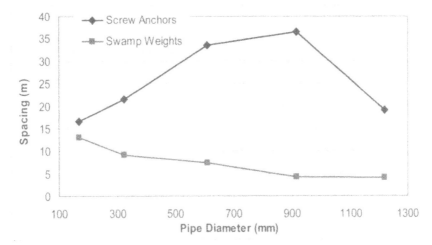

NOTES:

1. Pipe properties are based on typical values for a 10,000 kPa pressure pipeline.
2. Swamp weight spacing does include a benefit from shear between the swamp weights, anchor spacing does not include any benefit from soil shear.
3. Anchor spacing is to nearest 0.5 m and includes a temperature differential of 50 °C, a maximum anchor working load of 75 kN, and a maximum pipeline uplift between anchors of 100 mm.

Figure 23: Swamp Weight vs. Screw Anchor Spacing Comparison

5.7.5   Buoyancy Control Item Locations Work Process

The typical work cycle for buoyancy control on a project consists of:

221

1. Assessment of buoyancy control needs

2. Application of design principles of the various available buoyancy control options, and summing of the total quantities of each option

3. Procurement of the needed materials and obtaining installation prices via the construction bid and award process (maximum flexibility in material delivery will aid in reducing final costs)

4. Installation of buoyancy control materials based on the results of Steps 1 and 2, but to varying degrees, taking into account the actual field conditions

5. Recording accurate as-built information

The first phase, assessment, can be a simple step, or it can be a step that is done several times through a project. Optimally, from a cost effectiveness perspective, overland buoyancy control is placed where it is required, i.e. in areas where the depth of organic soils exceeds the calculated allowable depth. This approach generally requires cycling through the assessment phase each time additional data becomes available. An alternative approach would be to use buoyancy control in all suspect areas. This approach can reduce the assessment phase to one simple step. Although this approach may reduce risk, assuming all suspect areas are properly identified, it generally increases construction costs.

Additional details on steps 1 to 5 follow.

5.7.5.1 Location Assessment Phase I

As previously noted, the phase where buoyancy control needs are assessed can be a single step or it can be multiple steps. In the early stages of a project, the level of office and field efforts needed to assess right-of-way soil conditions and quantify buoyancy control requirements should be compared to the potential overall project's buoyancy costs. Potential quantities should reflect project or owner acceptable risk levels of post-construction pipe exposure or pipe flotation. Generally, if potential buoyancy costs are high compared to the chosen risk level, an increased assessment effort is justified.

The first cut at weighting requirements for a project can be done using a relatively inexpensive desktop analysis. For greenfield projects, this can consist of a terrain analysis of stereo aerial photography or satellite photography. (Please refer to the chapter on terrain analysis.) The analysis would consist of identifying likely areas of organic deposits. An experienced air photo interpreter could also estimate the potential depth of the organics. This would provide initial quantities for various control options and length requirements for the overland portions of the pipeline. For looping projects, the first cut assessment can be as simple as specifying buoyancy control at the same locations used on the original pipeline. A possible 'correction' factor can be applied if the pipe is a different diameter, and either less or more buoyant than the original pipeline.

5.7.5.2   Location Assessment Phase II

The next assessment level, field investigations, requires substantially greater effort and funds. The amount of each will depend greatly on the timing of the assessment and the ability to access the proposed right-of-way. Right-of-way accessibility generally depends on land ownership and the project's location relative to existing infrastructure. Remote projects are of course more difficult and costly to access. Land ownership can affect accessibility in part on whether it is private or government owned land. In many cases, access to government land is quicker to obtain than to private land. Special land designations such as nature reserves can be an exception and may be particularly difficult to access. The amount of effort and funds required will also depend on the type of planned fieldwork.

Fieldwork can consist of:

1.   continuous geophysics (including techniques such as ground penetrating radar (GPR), fixed or variable frequency electromagnetics, seismic, or electrical resistivity imaging (ERI))

2.   borehole drilling and/or test pit excavation

3.   a combination of 1 and 2.

Option 1 has the advantage of needing a minimal of access and ground preparation. It can be done using small motorized equipment such as quads or snowmobiles. It can be done in many cases on foot only, thus further reducing the clearing requirements or environmental impact. A decision on the type(s) of geophysical technique to use is best left to a geophysicist experienced in this type of work. The geophysicist also should understand how the data is being applied in the design as this may affect the technique(s) used. Additional background information is available in Thomas and Henderson and Henderson et al.

Sufficient lead time before construction is required for geophysics. The data interpretation for a longer pipeline can take several weeks. This needs to be followed by some time for the designer to review the data and incorporate the results into the design.

Option 2 can be done with hand equipment, or with a small drill mounted on a quad. Hand equipment can consist of a small probe pushed through the muskeg until mineral soil is encountered or to the probe's maximum penetration depth. Hand equipment can also consist of a generator (carried in a small trailer pulled by a quad) powered hand drill. The drill is advanced until either competent ground is encountered or to the drill's maximum penetration depth. Option 2 may also involve truck or track mounted drills, track mounted excavation equipment (backhoes) or a heli-portable drill. Ground based equipment necessitates extensive clearing in wooded areas and access preparation, if access can even be cost-effectively prepared. The optimal timing of larger ground based equipment is thus

223

following initial access preparation just ahead of pipeline construction. Heli-portable drills reduce access needs, but helicopter time generally is expensive.

Option 3 is generally the preferred option for a longer pipeline. Properly interpreted geophysics used in combination with boreholes can provide a more complete picture of the subsurface conditions than either item on their own. Geophysical interpretation can be subject to error unless there are boreholes drilled to allow proper calibration of the geophysical readings to the soil types. Borehole drilling alone may miss transition zones or undulations in the depth of the organic layer, even if a large number of holes are drilled. Geophysics can reduce the required number of boreholes, while the boreholes provide the data needed to optimize the interpretation of the geophysical readings.

The timing of this phase, as previously noted, is subject to access and cost in combination with when the data 'must' be available to the designer. If the designer can wait until the early phases of construction when access has been opened up, the investigation costs generally will lower than if the work is done ahead of construction, but keeping mind the interpretation time required for geophysics. The timing will in part depend on the planned type of buoyancy control, as well as on the procurement philosophy.

5.7.5.3   Location Assessment Phase III

A phase III assessment could be defined as the inspection of the excavated trench. This will provide accurate data on organic thickness, but will not provide information on soil conditions below the trench base. This may affect the applicability of anchors. It also greatly restricts the ability to adjust the buoyancy control based on the subsurface conditions. It does not allow for using extra depth trench, unless substantial costs are incurred by bringing equipment back to re-dig the trench. Also, if swamp or river weights have already been laid along the RoW or bolted to the pipeline, it is expensive to remove the weights and change to an alternate method of buoyancy control. In these situations, the weights are usually installed even if they are not required, or are not the optimal type of buoyancy control, unless it is suspected the project may run short.

An assessment of ditch soils can be useful when anchors are the specified buoyancy control. It is easy to delete or add anchors as required. Their installation does depend though on whether the subsurface soils are conducive to anchors.

5.7.6   Procurement

The preferred procurement process for buoyancy control will depend on the assessment phase(s). If only a single phase assessment is done, all the estimated buoyancy needs are ordered ahead of construction. The construction drawings and bid documents will show the proposed quantities.

If the assessment process includes a second level, the procurement process likely also has a second phase. This will depend in part on the lead time required for ordering the buoyancy control materials proposed for the particular project. Most buoyancy control material does require a minimum 2-3 month lead time as only nominal inventory quantities are generally carried by suppliers. Anchors do have an advantage in that some parts of the anchor assembly, such as the extensions and couplings, and to a lesser degree the lead anchor, are not pipe diameter specific. This provides additional flexibility in stocking and re-stocking of returned material. Similarly, granular bag weights, although pipe diameter specific, do not all have to be filled well ahead of construction. A reserve can be kept aside and the bags filled only once it is known they are needed. Unfilled bags may be returned to the supplier fairly easily, depending on the contract arrangement. Alternatively, the prime contractor or the owner can store some material in their own warehouse for use on other projects.

For a multi-phase assessment, an initial buoyancy order may be submitted based in part on the first phase assessment. The order though, is made with the understanding that changes are expected. Suppliers generally would be asked to include the costs associated with and the lead time required to order and ship additional material. The buoyancy designer must keep in mind that the shipping location for the supplier may only be an intermediate destination. Additional time may be required to ship from the supplier's delivery point to the pipeline right-of-way.

Similarly, construction bid documents to the prime pipeline contractor may go out with preliminary quantities, and the contractor would be asked to provide add and delete prices for each type of buoyancy control option that is proposed for the project. Addendums to procurement and construction bids would be issued as needed as additional buoyancy quantity assessments were completed.

5.7.7 Installation

The buoyancy materials generally will be installed as shown if only a single assessment phase was completed. The construction supervisor may need to supplement ordered materials with extras if site conditions result in additional areas being identified that require buoyancy control.

For a multi phase buoyancy control quantity assessment, construction may proceed in the same manner as just described if the assessment did not include any work following the issuing of the prime contract bid documents. If however, the assessment includes such items as boreholes or test pits following access preparation by the prime contractor, or inspection of organic layer thickness in the excavated pipeline trench, then "just in time" buoyancy control design and installation would be part of the construction process. The contractor would be prepared to adjust ditch depth, or place concrete or gravel weights in the required locations only shortly before actual installation. Buoyancy control requirements must, however, not be too late so as to disrupt the overall construction flow, as this may result in significant extra charges being submitted by the prime contractor. These charges would be over and above the 'add and delete' prices that would be part of the contract.

5.7.8   As-Built Drawings

An important part of any project is the compilation of as-built records and drawings. These records not only provide a means to finalize payments to the contractor, they also provide valuable data for other projects that may be built alongside, or nearby to the completed project. Accurate records can result in significant savings in future field investigations and design efforts.

5.8   References

1. A. B. Chance Co., Pipeline Anchoring Encyclopedia

2. CSA Z662-07 - Oil and Gas Pipeline Systems.

3. G. Connors, The Use of Geotextile Fabric, Anti-Buoyancy Weights for Buried Pipelines, Paper IPC04-0028, Proceedings of International Pipeline Conference, Calgary, Alberta, Canada, October 4-8, 2004

4. S. L. Couperthwaite and R.G. Marshall, Design Methodology for Buoyancy Control in Muskeg Terrain: Recent Developments, American Society of Mechanical Engineers, Pipeline Engineering Symposium, Vol 6, 1987

5. R. Doering and R. Robertson, Screw Anchor Buoyancy Control Saves an Estimated $12 Million for Enbridge Pipeline

6. EBA Engineering Consultants Ltd., Installation and Testing of Rod Anchors in Permafrost at Norman Wells, N.W.T., Prepared for Canadian Arctic Gas Study Limited, January 1973 & June 1974 (revised)

7. J. Henderson, M. Bowman & J. Morrisey, The Geophysical Toolbox: A Practical Approach to Pipeline Design and Construction, Paper IPC04-0190, Proceedings of International Pipeline Conference, Calgary, Alberta, Canada, October 4-8, 2004

8. M.P. Mitsch, and S.P. Clemence, The Uplift Capacity of Helical Anchors in Sand, ASCE, Uplift Behaviour of Anchor Foundations in Soil, pp 26-47, 1985

9. J. S. Mooney, S. Adamczak, and S.P. Clemence, The Uplift Capacity of Helical Anchors in Clay and Silt, ASCE, Uplift Behaviour of Anchor Foundations in Soil, pp 48-72, 1985

10. R. Robertson and R. Curle, Buoyancy Control of Large Diameter Pipelines in Canada Using Helical Screw Anchors; PD-Volume 69, Pipeline Engineering, ASME 1995

11. G.R. Simmonds and L. N. Thomas, Incorporating Muskeg Soil Shear Strength into Buoyancy Control Design

12. L.B. Smith and N. Hernadi, The Design and Construction of Grouted Pipe Piles in the Canadian Arctic, American Concrete Institute Seminar, March 14 and 15, 1989

13. L. Thomas & J. Henderson, An Integrated Approach to Pipeline Buoyancy Control and Implementation, Paper IPC04-0429, Proceedings of International Pipeline Conference, Calgary, Alberta, Canada, October 4-8, 2004

6. Geohazard Management

6.1 Introduction

Why is this an increasingly important topic?

The suite of activities constituting geohazard management practice has seen considerable advancements in recent years as the topic has gained prominence within the pipeline community. It is interesting to take stock as to why this topic is becoming increasingly important. At the time of this book's publication, there are at least five major drivers contributing to this topic's advancement which will be briefly discussed, namely:

1. A better understanding of the prevalence and frequency of these hazards
2. A range of business drivers promoting proactive management of these hazards
3. The availability of new enabling tools as a matter of state of practice
4. Worldwide pipeline expansion in areas of difficult terrain
5. An active peer community contributing to this evolving specialized topic

Several publications provide a sense of the prevalence and frequency of geohazards. Table 1 summarizes reported pipeline failure frequencies associated with geohazards from various areas around the world. Caution is required in reviewing and applying these general statistics given inconsistent recording of pipeline failure causes and in the calculation methods of of reported frequencies.

Table 1 - Reported pipeline failure frequencies associated with geohazards world-wide

Geographic Areas	Reported Pipeline Failure Frequencies Associated with Geohazards (per 1000km-yr)	Pipeline Product	Reference Document
USA	4.5×10^{-3}	Gas	United States Department of Transportation – OPS (for the period between 1984 – 2001)
Europe	1×10^{-2}	Gas	Sweeney, 2005
Canada – Province Alberta	1.6×10^{-3}	Gas	Alberta Energy Utilities Board website (for the period between 1980 to 1997)
	3.2×10^{-3}	Oil	
Canada	4.22×10^{-3}	Gas	National Energy Board website (for the period between 1984 to 2003)
	5.43×10^{-3}	Liquid	
South America - rugged	2.8 (older) to 0.33 (new pipelines)	Both	Sweeney, 2005

Generally, geohazard induced pipeline failures occur in areas known by both the local community of geotechnical specialists and the pipeline operators to be prone for certain mechanisms to act. Depending on geomorphologic conditions and pipeline alignments, these hazards may act in very few localized areas along a pipeline corridor while in other cases may have a nearly continuous spatial distribution. Statistically, in contrast to metal loss or mechanical damage induced failures, geohazard related pipeline failures are typically decidedly less frequent in Europe and North America. Still, even where they may be relatively less frequent, they are typically associated with higher consequence.

Increasingly, the practices of prudent pipeline owners include proactive management of this class of hazards in response to business drivers. These best practices are due to at least 2 factors, namely;

1. There are heightened expectations of pipeline companies from the external stakeholders including regulators and the public at large. Ensuring public safety and accepting a role of environmental stewardship are now accepted as integral parts business. The avoidance of ruptures and leaks in valleys,

230

as an example, where several geohazards may be concentrated and may be in a cause and effect inter-relationship, become particularly important due to the associated consequences.

2. Financially, operators need to optimize internal expenditures associated with all aspects of pipeline integrity management. Prudent operators strive to move from reactive to proactive and predictive prevention for all hazards including geohazards. Increasingly, risk-based assessment and integrity management planning practices (as will be described later in this chapter) provide the basis of predictive prevention through a defendable methodology to support optimized expenditure profiles from year to year.

The availability of somewhat recently developed enabling tools and technologies for use in the state of practice now provides practitioners a step-change improvement in their ability to manage geohazards potentially impacting pipelines. What was new in the early nineteen nineties, GIS and database tools are now found in common use by pipeline companies small and large either in-house or through the supporting community of consultants. In addition, there is a growing suite of proven remote sensing methods capable of precise and detailed topographic characterization such as LiDAR or the detection and measurement of sub-centimeter ground movements such interferometery analysis of radar imagery. These and similar relatively recent technology advances enable pipeline operators and the community of geotechnical specialists to better manage geohazards.

World wide pipeline expansion in areas of difficult terrain has proved to be another driver of recent advances in geohazard management practice. After building and operating massive networks of pipelines in the relatively low and limited geohazard prone terrain in much of North America and Europe, pipeline development and expansion in areas of difficult terrain called for different approaches in design, construction and operations where geohazards management is by far the dominant consideration. A combination of rugged terrain, heavy precipitation and pervasive seismic activities presents itself to designers and operators in several South American pipeline developments. Different geohazard related design and operational demands are presented in both Arctic and desert pipeline developments.

An active peer community contributing to this evolving specialized topic at the time of this book's publication has been an asset to the rapid developments in this field. Recent years has seen significant contributions in the field of pipeline geohazard management reported in Pipeline Research Council International (PRCI) reports, American Lifeline Alliance documents, as well as the proceedings of the 2004 International Conference on: Terrain and Geohazard Challenges facing Onshore Oil and Gas Pipelines and numerous American Society of Mechanical Engineers (ASME) conferences. At the time of this book's publication, a guideline document was under development through the joint sponsorship of the PRCI and the United States Department of Transportation relating to pipeline design and operations in geohazard areas. Beyond a strictly pipelines context and more generally relating to geohazard risk assessment, the proceedings of the 2005 International Conference on Landslide Risk Management edited by Hungr, Fell, Couture and Eberhardt and the 2004 book titled "Landslide Risk Management" by Lee and Jones are both excellent reference documents that provide analytical methods that may be adapted into the pipeline context.. Also while beyond the strict pipelines context, regular publications and updates by geological surveys of various jurisdictions worldwide should be considered as valuable source to practitioners in the pipeline geohazard management field in identifying potential geohazards in an area of interest.

To recognize the breadth of the active peer community contributing to this filed, this chapter will note many related publications beyond those directly referenced.

What is included in the chapter?

As geohazard management starts with routing and design and continues through operations, this chapter's treatment of the subject will present methods for both design and operations. Figure 1 proposes a general flow chart for geohazard management parenthetically identifying the order of topics as they will be presented in the chapter.

Following the introduction, the second section of the chapter will be a discussion of potential regional geohazards will be presented including how they may interact as well how they impact pipeline integrity. The third section of the chapter will overview data management aspects that are so integral to best practices in geohazard management. An overview of risk assessment methodologies will be presented in the fourth section. An introductory discussion of the range of analytical methods and tools used to derive probabilities of failure is also included in the fourth section. The fifth and sixth sections of the chapter will provide overviews of the monitoring and mitigation toolboxes available to pipeline operators. Finally, a discussion of the considerations that effect the decision process undertaken by an operator in managing geohazards on a year to year basis is presented.

An important philosophical starting point for the chapter!

While mindful of corporate responsibilities to the environment in the pipeline's vicinity and as especially due to the pipeline's construction and operations, pipeline integrity relating to geohazards should be managed from the pipeline out as opposed to a purely geotechnical treatment of the hazard.

This important philosophical point drives a key technical consideration of incorporating the vulnerability or structural capacity of pipelines in estimating the probability of failure will be discussed in section 6.4.

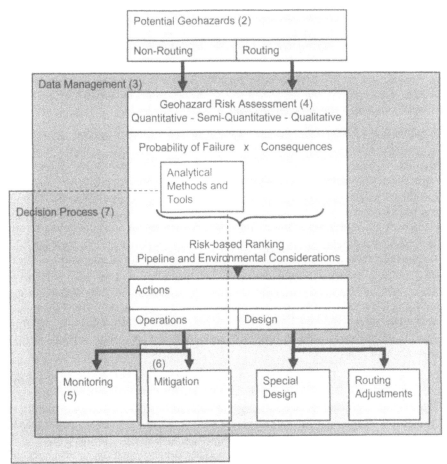

**Figure - 1 A simplified flow chart of pipeline geohazard management (#)
Chapter sub-sections addressing the respective topics identified in
parenthesis.**

6.2 Regional Geohazards

General

Oil and gas pipelines are linear components of distribution networks.   Major
pipelines may traverse long distances along a relatively narrow right-of-way,
crossing a variety of geologic and climatic environments.   Within these
environments, natural hazards of a geological, hydrological or tectonic origin
(termed geohazards) may constitute a subset of the potential threats to which the
pipeline is exposed.

The range and relevance of geohazards depends in part on the physiographic setting of the pipeline route. A number of common geohazards, such as slope instability, erosion, seismicity and river scour, are frequently encountered in temperate and other harsher environments. Pipelines in arctic/subarctic, equatorial, and desert environments may be face with additional challenges for design, construction and long-term operation of pipelines related to extreme climatic conditions and associated events.

In addition to physiographic setting, pipeline design specifications also determine, in part, the relevance of particular geohazards. For example, a high operating pressure pipeline design using thick-walled pipe is less susceptible to damage from some geohazards than a design incorporating thin-walled pipe. Design specifications are dependant on the anticipated loads and operating conditions associated with the pipeline, as well as the regulatory framework and standards in place for a given pipeline project. The operating conditions and anticipated strain demand on the pipeline influence the selection of pipe material (i.e., steel grade), welding techniques, pipe diameter, pipe wall thickness, pipe coating and a host of other design, construction and operation specifications. Estimating cumulative strain demand requires an understanding of potential loading mechanisms, including those imposed by geohazards, and other factors that may affect the long-term integrity of the pipeline. The inter-relations of physiographic setting, pipeline design specifications and geohazards must be taken into account in the assessment of risk.

Pipeline operators should recognize that geohazards have the potential to affect not only the condition of the pipe, but also the pipeline ditch and the right-of-way. These three elements of the pipeline system may each require expenditures through the operational life of the pipeline to maintain system integrity, reduce risk, and limit environmental effects. While pipe integrity is a primary driver for pipeline management expenditures, secondary drivers such as environmental sensitivity may introduce additional demands on the pipeline maintenance budget. For example, material transported off the right-of-way through some natural process (e.g., mass movement or erosion) may not affect pipe integrity, but may require remediation. These secondary drivers should be anticipated to the extent possible in planning a pipeline maintenance program.

For a given pipeline project, it is important to focus on credible probable geohazards within the pipeline corridor that may directly affect the three main pipeline elements (i.e., the pipe, the ditch, and the right-of-way). Some of these geohazards may constitute environmental loads and effects triggered by pipeline installation and/or operations. Others exert external loading directly on the pipeline. In some situations, multiple geohazards at a given location may act independently, may offset one another, or may reinforce one another in terms of effects on the pipeline elements. Geohazard assessment of the pipeline route is therefore aimed at applying a rational, systematic approach to identify where

234

geohazards may occur, where multiple geohazards may overlap, and what the relative effects of these independent and combined geohazards may be on the three main pipeline elements.

Geohazard Inventory

The first step in preparing for geohazard assessment is the development of a geohazard inventory (Rizkalla, 2007). Table 2 provides a comprehensive list of geohazards. Ten general geohazard categories associated with one or more of the challenging geographic environments considered in this section are defined. These include the following:

1. Landslides/mass movement – phenomena associated with movement of a large volume of soil, rock, or snow/ice.
2. Tectonics/seismicity – phenomena associated with seismic activity and fault movement.
3. Hydrotechnics – phenomena associated with watercourse and waterbody hydrodynamics.
4. Erosion – phenomena associated with the transport of soil particles by surface water, groundwater or wind.
5. Geochemical – phenomena associated with soil, rock, or groundwater chemistry.
6. Freezing of unfrozen ground – phenomena associated with freezing of soil and groundwater.
7. Thawing of permafrost – phenomena associated with thawing of frozen soil and ground ice.
8. Unique soil structure – phenomena associated with potentially detrimental soil characteristics.
9. Desert and sand sea geotechnics – phenomena associated with unique desert conditions.
10. Volcanic activity – phenomena associated with active volcanoes.

There are 40 geohazards identified in Table 2 under these ten categories that serve as the basis for an initial geohazard screening assessment. From an engineering treatment viewpoint, some of these geohazards can be further subdivided on the basis of other criteria such as longitudinal versus transverse slope. For each geohazard, the phenomenon and associated effects are described, and the potentially affected pipeline elements are identified.

Additional geohazards such as isostatic and valley rebound may constitute additional phenomena to be considered under some circumstances, but these processes typically occur slowly over a long time period. Other unique geohazards (e.g., normally occurring radioactive materials) may also exist at some locations, and can be added to Table 2 as necessary to develop a complete inventory for a specific project. Finally other considerations that require geotechnical review and input such as the assessment of blasting and surface loads should not be overlooked during pipeline design and operations. That PRCI's wealth of publications offer practitioners state of practice references for these and other topics.

Table 2 also provides a stylized checklist to identify options for mitigating the identified geohazards. These options include:

- Analysis - specialized analysis to improve quantification of the likelihood of occurrence of, or risk associated with, a particular geohazard.
- Field - field investigation to gather characterization data related to a geohazard.
- Design - design solutions to reduce the potential effects of a geohazard on the pipeline system to an acceptable level.
- Monitoring - monitoring to track the progression of a particular geohazard and its associated effects on the pipeline system.
- Maintenance – regular routine activities to counter the effects of a geohazard on the pipeline system.

Table 2 - Inventory of Geohazard Phenomena and Effects

Common Occurrence	Geohazard	Description of Phenomenon and Potential Effects	Affected Elements			Mitigation Options				
			Pipe	Ditch	ROW	Analysis	Field	Design	Monitoring	Maintenance
	1. Landslides/Mass Movement									
✓	Deep-seated landslide	Deep rotational or complex failure of steep longitudinal or cross slopes → Rapid loading and deformation of pipe; exposure of pipe, or deeper burial of pipe; disturbance of ditch and ROW; pipe vulnerability depends on loading direction	X	X	X		✓	✓	✓	✓
✓	Slope creep	Gradual downslope mass movement of soil on frozen or unfrozen slopes → Gradual loading and deformation of pipe (longitudinally or laterally)	X	X	X		✓	✓	✓	✓
	Slope creep rupture	Sudden rupture of warm permafrost on slope following gradual downslope mass movement → Gradual loading and deformation of pipe leading to rapid loading and deformation (longitudinally or laterally)	X	X	X	✓	✓	✓		
	Thawed layer detachment	Shallow slope instability involving downslope movement of thawed soil layer above permafrost → Reduced cover and exposure of pipe; possible longitudinal or lateral loading on slopes with deep thawed layer	X	X	X	✓	✓	✓	✓	✓
✓	Rockfall or rock avalanche	Downslope movement of either individual blocks or a disaggregated mass of rock → Sudden vertical loading of pipe and increased burial depth; possible obstruction of ROW and ditch; possible exposure of pipe	X	O	O			✓	✓	✓
✓	Debris flow	Downslope movement of water-saturated debris → Sudden vertical loading of pipe and increased burial depth; possible obstruction of ROW and ditch; possible exposure of pipe	X	O	O		✓	✓	✓	✓
	Snow avalanche	Downslope movement of either individual snow, ice and debris → Sudden vertical loading of pipe and increased burial depth; possible obstruction of ROW and ditch; possible exposure of pipe	X	O	O			✓	✓	✓

237

Table 2 - Continued

Common Occurrence	Geohazard	Description of Phenomenon and Potential Effects	Pipe	Ditch	ROW	Analysis	Field	Design	Monitoring	Maintenance
	2. Tectonics/seismicity									
✓	Fault displacement (fault crossings)	Movement along existing faults → Shear displacement/loading of pipe	X			✓	✓	✓	✓	✓
✓	Dynamic liquefaction	Sudden loss of strength and/or movement of soil subjected to dynamic loading → Lateral spreading of soil on ROW, pipe uplift (buoyancy), or pipe settlement leading to flexural strain and/or exposure	X	X	X		✓	✓		
✓	Dynamic ground motion	Ground shaking due to seismic loading → Dynamic loading of pipe	X			✓				
	3. Hydrotechnics									
✓	Vertical scour (water crossings)	Vertical hydraulic erosion of material over pipe in existing channel → Reduction of cover or pipe exposure; possible unsupported pipe span with vortex shedding	X	X				✓	✓	✓
✓	Channel migration (water crossings)	Migration of channel (lateral scour) in flood plain causing erosion of soil over pipe beyond the deep burial portion of the pipeline → Reduction of cover or pipe exposure; possible unsupported span leading to pipe strain aggravated by debris caught on pipe; possible toe erosion of valley slopes	X	X	X			✓	✓	✓
✓	Buoyancy (pipe uplift)	Low soil mass with elevated groundwater table or failure of mechanical buoyancy control leading to pipe uplift → Potential exposure of pipe; pipe flexure due to upheaval displacement in unrestrained pipe section	X	O			✓	✓	✓	✓
	Rapid lake drainage	Breach in lake impoundment causing rapid drainage of water and associated erosion along drainage path across ROW → Potential exposure of pipe and/or an unsupported span; vortex shedding aggravated by debris caught on pipe in extreme case	O	X	X	✓		✓	✓	✓

Table 2 – Continued

Common Occurrence	Geohazard	Description of Phenomenon and Potential Effects	Affected Elements			Mitigation Options				
			Pipe	Ditch	ROW	Analysis	Field	Design	Monitoring	Maintenance
	Coastal inundation and flooding	Change in sea level in low-lying coastal areas or water level of inland waterbodies → Inundation of ROW, buoyancy effects on pipeline, possible erosion due to wave action (tsunami)	X	X	X		✓	✓	✓	✓
4. Erosion										
✓	Backfill erosion (upheaval disp.)	Hydraulic transport of ditch backfill and associated loss of restraint on pipe → Pipe flexure/exposure due to upheaval displacement of pipe; exacerbated by soil warming/strength loss	O	X			✓	✓	✓	✓
✓	ROW erosion	Hydraulic transport of soil particles along ROW → Loss of soil cover on ROW; possible stream siltation			X		✓	✓	✓	✓
	Subsurface (piping) erosion	Hydraulic transport of subsurface material through groundwater flow → Loss of support beneath pipe; possible void development, possible stream siltation	X	O	O		✓	✓	✓	✓
5. Geochemical										
	Acid rock drainage	Attack on pipe by acid groundwater → Localized metal loss and possible reduction in structural capacity	X			✓		✓		
✓	Karst collapse	Ground subsidence due to sinkhole development (collapse of active karst or paleokarst solution cavities) below pipe → Potential for pipe exposure and unsupported pipe span depending on size and depth of collapse feature	X	O	O	✓	✓	✓		
	Saline soil/bedrock	Attack on pipe by saline groundwater → Localized metal loss and possible reduction in structural capacity	X				✓	✓		
6. Freezing of unfrozen ground										
	Frost heave (pipe)	Pipe uplift from frost bulb development in unfrozen spans → Pipe strain caused by flexure/uplift resistance of adjacent frozen soil	X			✓	✓		✓	✓

239

Table 2 – Continued

Common Occurrence	Geohazard	Description of Phenomenon and Potential Effects	Affected Elements			Mitigation Options				
			Pipe	Ditch	ROW	Analysis	Field	Design	Monitoring	Maintenance
	Frost heave (ditch)	Surface heave over ditch due to ground freezing → Disruption of surface drainage across ROW due to elevated ditchline		X	O				✓	✓
	Frost bulb development (cross-country)	Reduction of soil hydraulic conductivity due to ground freezing → Disruption of subsurface drainage beneath and above pipeline; possible increased pore pressure and aufice on side-slopes with active groundwater systems		X	O				✓	✓
	Frost bulb development (water-crossing)	Reduction of soil hydraulic conductivity and ground heave due to freezing; constriction of water flow path → Disruption of watercourse flow and fish habitat or fish migration; intervention during operation potentially disruptive to watercourse		X	O	✓	✓	✓	✓	✓
	Frost blister	Ground heave due to increased hydraulic pressure beneath frozen layer from subsurface drainage disruption on cross slopes → Elevated pore pressure and possible erosion associated with release of trapped water or melting of ice upslope of pipeline		O	X				✓	✓
	Ice-wedge cracking	Sudden extensional cracking of ice-wedges in frozen ground → Generation of localized extensional displacement/tensile stress in pipe, and associated shear stress near location of ice-wedge cracking	X			✓				
7. Thawing of permafrost terrain										
	Thaw settlement (pipe)	Differential settlement of the pipe due to thawing of permafrost → Pipe strain caused by pipe flexure in areas of differential settlement in frozen ground, or at frozen/unfrozen interfaces	X			✓	✓		✓	✓
	Ditch settlement	Loss of pipe cover and soil strength due to thawing of ice-rich backfill → Increased potential for erosion along sunken ditch, pipe exposure, and/or upheaval displacement of pipe	O	X	O	✓	✓	✓	✓	✓

240

Table 2 - Continued

Common Occurrence	Geohazard	Description of Phenomenon and Potential Effects	Affected Elements			Mitigation Options				
			Pipe	Ditch	ROW	Analysis	Field	Design	Monitoring	Maintenance
	Thaw settlement (ROW)	Subsidence along ROW due to thawing of permafrost → Disruption of surface drainage across ROW due to ground subsidence			X			✓	✓	✓
	Thaw bulb (slopes)	Rapid thawing of ice-rich permafrost on slopes → Pore pressure generation and loss of soil strength leading to thermal erosion; possible void formation beneath pipe; contributing factor to slope movement	O	X	X				✓	✓
	Thermokarsting/massive ice	Thawing of massive ice or ice wedges leading to surface subsidence and possible ponding/sustained thawing → Disruption of surface drainage across ROW due to subsidence, and possible buoyancy issues	O	O	X		✓	✓	✓	✓
8. Unique Soil Structure										
✓	Boulder/cobble/rock indentation	Interaction of pipe with cobbles/boulders or shallow rock at bottom of ditch → Point loading/denting of pipe	X				✓	✓	✓	✓
	Static liquefaction	Collapse of sensitive (flocculated) soil structure due to surface loading or groundwater effects → Disruption to right-of-way; potential for pipe exposure	O	O	X		✓			✓
	Sensitive and residual soil	Disturbance to sensitive or residual soils during pipeline or access construction → Rapid loss of soil strength leading to flow or ground instability on ROW possibly inducing strain in pipe due to lateral loading, deeper burial, or loss of support.	X	X	X		✓	✓	✓	✓

241

Table 2 – Continued

Common Occurrence	Geohazard	Description of Phenomenon and Potential Effects	Affected Elements			Mitigation Options				
			Pipe	Ditch	ROW	Analysis	Field	Design	Monitoring	Maintenance
	9. Desert and Sand Sea Geotechnics									
	Dune migration	Wind transported desert soil → Change in burial depth resulting in increased or decreased load on pipe, possible pipe exposure, possible deeper burial.	X	X	X		✓	✓	✓	✓
	Flash flooding/scour at wadis	Water transported desert soil → Change in burial depth resulting in decreased load on pipe, possible pipe exposure.	X	X	X			✓	✓	✓
	10. Volcanic Activity									
	Ash falls	Accumulation of ash → Change in burial depth resulting in increased load on pipe, obstructed access to pipeline	X	O	O				✓	✓
	Lahars	Mudflow comprising pyroclastic material and water → Change in burial depth resulting in increased or decreased load on pipe, obstructed access to pipeline	X	X	X			✓	✓	✓
	Pyroclastic flow	Accumulation and flow of pyroclastic material→ Change in burial depth resulting in increased or decreased load on pipe, possible pipe exposure, obstructed access to pipeline	X	X	X				✓	✓

X – Direct effect on pipeline element (pipe, ditch, right-of-way)

O – Indirect or secondary effect on pipeline element

The suitability and effectiveness of these options will vary from project to project depending on a number of factors including cost, availability, access, and schedule related to each mitigation option. Some geohazards may be best mitigated by avoidance at the design stage through route refinements within a specified corridor where feasible. Geohazards that fall under this category may include steep unstable slopes, karst features, active faults, unfavourable watercourse crossing locations, potential flood areas, permafrost with high ice content or massive ice, unusually sensitive soil deposits, and active pyroclastic flow areas. For geohazards not avoided during design, a mitigation strategy that includes the full range of options listed above provides flexibility in optimizing mitigation costs/benefits, and prioritizing sites for mitigative actions in near-term and future operations/maintenance budget cycles.

For a given pipeline project, the geohazards in Table 2 should be reviewed, and the list reduced to fewer credible probable geohazards that are known or expected to occur along the pipeline route. The revised list then becomes the basis for subsequent geohazard assessment.

Geohazard Description

A detailed description of each geohazard is required for geohazard assessment. A standard template is used to capture key information required to identify each geohazard and to estimate the relative susceptibility of terrain along the pipeline route to the geohazard. Once completed for each geohazard, the suite of templates is maintained throughout the project life to ensure consistent characterization of geohazards. The templates are updated as new information becomes available to refine the geohazard assessment (e.g., replacing conservative assumptions with route or operational data).

Typical information captured by the geohazard summary template, based on a particular semi-quantitative index-based assessment approach (NEB, 2006), is as follows:

1. Geohazard Name – from the geohazard inventory
2. Process Description - process/event associated with the geohazard
3. Process Rate - rate at which the geohazard develops
4. Potential Effects – potential effects on the pipe and on other pipeline elements
5. Triggering Mechanisms - mechanisms that may initiate the primary geohazard
6. Controlling Parameters – primary and secondary attributes controlling geohazard occurrence
7. Contributing Factors - additional attributes that may influence the geohazard

8. Relation to Other Geohazards - possible linkages to other geohazards, including load combinations and trigger/event pairs
9. Constraints - spatial constraints on the area of influence of the geohazard
10. Identification Criteria – specific criteria required to identify the geohazard
11. Ranking Criteria – specific criteria for ranking susceptibility associated with the geohazard
12. Mitigation Options – options to reduce susceptibility and effects on index values
13. Linkage to Field Investigation - information required from field investigation to improve identification or ranking of the geohazard
14. Data Sources – GIS data sources required to conduct the geohazard assessment
15. Query – GIS query of data sources
16. Specific Output - specific features and data layers required to show on maps and as 3D images to delineate area of susceptibility associated with the geohazard
17. General Comments - pertinent comments on refinements to the template or other information that may help in geohazard assessment.

Specific details related to some of these information categories (e.g., ranking criteria) will vary depending on the risk assessment approach adopted for a project. Geohazard risk assessment is discussed further in the next section.

6.3 Data Management

At all stages of a pipeline lifecycle, spatial information plays a significant role. From archived public aerial photographs and value-added derivatives such as Digital Elevation Model (DEM) products and terrain mapping, through to higher resolution imagery and LiDAR, data is collected and utilized. Data bases and GIS tools are essential to manage the great volumes of raw data and derived information. Certainly some of these tools have been long available and in use by the pipeline industry. Still, the relatively recent availability and the power of additional data sources and data management capabilities have been a real boon for advancing the state of practice of geohazard management. Today, those involved in an operating pipeline company's geohazard management, may they be in-house or external consultants, can ill-afford to remain outside this arena.

Pipeline Design and Integrity Management Data Requirements and Sources

Considering the more commonly encountered slope instability related and hydrotechnical geohazards, the specific data requirements and the typical data

244

sources to meet these requirements are considerable. A summary of the data requirements and potential sources is presented in Table 3.

Among the data sources indicated in Table 3, is subsurface information derived from borehole drilling and test pit programs followed by laboratory testing. The deployment of aerial and ground-based geophysical surveying techniques may be a valuable complimentary source of subsurface data. There is a wide range of design, construction planning and integrity management drivers during operations that call for undertaking what may in some cases be difficult and costly field programs. However, once design and construction planning field program data is acquired and if properly retained in query-suited GIS-enabled data bases, it may serve as a valuable starting point for data requirements during operations. Among the drivers for undertaking these programs are:

- Slope stability analyses throughout the lifecycle of a pipeline
- River crossing designs and construction planning
- Horizontal directional drilling designs and construction planning
- Complex highway and other pipeline crossing designs and construction planning
- Foundation design for valves and in-line inspection launchers and receivers
- Upheaval instability assessment and mitigation designs
- Erosion control design optimization
- Design mitigation for liquefaction
- Seismic fault characterization
- Significant cut slopes (especially in geo-thermally sensitive terrain)
- Ditching methods confirmation
- Ditch import backfill, bedding and padding quantities
- Buoyancy control design optimization

Another particularly important class of data as indicated in Table 3 is topographic information. Digital elevation models (DEMs) are the computerized format of point elevation data representing the earth surface. The long available ortho-rectified photos and the relatively more recent LiDAR acquisition are the data sources that may be processed to generate the DEMs and contour mapping. LiDAR is the acronym for the laser pulses based "Light Detection and Ranging" technology. The data acquired in a LiDAR over-flight mission is a dense cloud of 3D points (X, Y and Z coordinates) that may be analyzed to produce a bare-earth DEM.

Table 3 - Geohazard Management Data Requirements and Sources

Information Required to Support Geohazard Assessments	Source of Data																
	Photo-based Digital Elevation Model	Terrain maps/reports	Topographic maps	Air photos	Satellite images	Pipe book	Alignment sheets (as-built)	Watercourse studies	Seismic studies	Climate data	Construction reports	Lab testing reports	Mitigation reports	Monitoring reports	LiDAR survey for new DEM	InSAR monitoring of route	Geophysics of route
Slope Instability Data																	
- Slope angle	X		X						X				X		X		
- Slope height	X		X										X		X		
- Slope length	X		X										X		X		
- Slope aspect	X		X										X		X		
- Vegetative cover		X			X								X				
- Failure features (cracks, scarps, bulging)			X										X			X	
- Failure/deformation mode			X		X				X				X	X		X	
- Rate of down slope movement										X	X	X	X	X		X	
- Rate of retrogression													X	X		X	
- Soil stratigraphy		X										X		X			X

246

Table 3 – Continued

Information Required to Support Geohazard Assessments	Source of Data															
	Photo-based Digital Elevation Model	Terrain maps/reports	Topographic maps	Air photos	Satellite images	Pipe book Alignment sheets (as-built)	Watercourse studies	Seismic studies	Climate data	Construction reports	Lab testing reports	Mitigation reports	Monitoring reports	LiDAR survey for new DEM	InSAR monitoring of route	Geophysics of route
- Depth to bedrock	X	X											X			X
- Depth of failure surface	X												X			X
- Distance from slope toe to watercourse	X		X		X											
- Slope width parallel to ROW	X											X		X		
- Minimum distance from pipe centreline to slope	X				X							X		X		
- Position of slope relative to pipe	X				X							X		X		
- Distance from slope toe to watercourse	X				X											
- Location of debris flow gullies	X				X		X					X				
Pipeline Mechanical Properties						X										
Watercourse Conditions																
- Crossing locations	X	X	X	X	X		X			X		X				

247

Table 3 – Continued

Source of Data

Information Required to Support Geohazard Assessments

Information Required	Photo-based Digital Elevation Model	Terrain maps/reports	Topographic maps	Air photos	Satellite images	Pipe book	Alignment sheets (as-built)	Watercourse studies	Seismic studies	Climate data	Construction reports	Lab testing reports	Mitigation reports	Monitoring reports	LiDAR survey for new DEM	InSAR monitoring of route	Geophysics of route
- Crossing method								×			×						
- Depth of cover							×	×			×						
- Channel type	×			×	×			×									
- Channel width	×		×	×	×		×	×									
- Stream type	×			×	×			×									
- Bed material type								×				×					
- Stream velocity (normal and flood)								×		×							
- Stream flow volume (normal and flood)								×		×							
- Potential upstream damming	×																
- Potential downstream damming	×																
- Weight locations							×	×			×		×				

248

Based on standards provided by the US Federal Geographic Data Committee (Federal Geographic Data Committee, 1998), the horizontal and vertical accuracies of topographic data produced from various raw data sources are presented in Table 4

Table 4 - Topographic Data Accuracy Derived from Various Raw Data Sources

Raw Data Source	Vertical Accuracy	Horizontal Accuracy
1:30,000 Aerial Photos	+/- 2.5 – 3.0 m	+/- 3.0 m
1:10,000 Aerial Photos	+/- 0.6 – 0.8 m	+/- 1.2 m
LiDAR	+/- 0.25 m	+/- 0.5m

Examples of Data Management Systems

With the accumulation of a wide range of spatial and site-specific data, the effective management of what is in fact a valuable asset to the pipeline operator in the form of these data sets becomes an important consideration. The results of regular inspections and ongoing monitoring programs, add to the accumulating body of information that are essential in managing geohazards acting on pipelines. Discussions of both monitoring techniques and the use of data in geohazard management planning decisions follow later in this chapter.

The main attributes of effective data management in the present context include:

1. effective and intuitive communication of the available data in various formats to stakeholders
2. accessible databases within a pipeline operator's organization
3. linkage to risk based decision modules
4. a well defined and managed information technology process to assure quality and security

In recent years, the description of mature geohazard data management systems has been reported in technical publications. Both proprietary in-house systems such as that reported by Petrobras (da Rocha, 2004) and commercially available data management systems have been reported (Leir, 2002; Leir, 2004, Greaves. 2007, and Reed, 2008). Other generally similar data management systems have been reported. These systems provide examples of what may be anticipated to become the state of practice for prudent pipeline operators large and small dealing with geohazards.

Figures 2, 3 and 4 provide examples drawn form the two reported of the effective and intuitive communication of data available for risk assessment analyses.

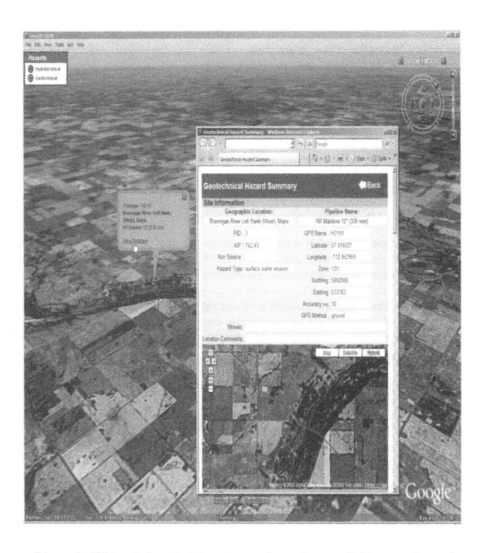

Figure 2 - With a linkage to internet mapping software, clicking on a hazard symbol displays a summary of the hazard site. (Published with permission from BGC Engineering Inc. as an update to figures presented in (Leir, 2004))

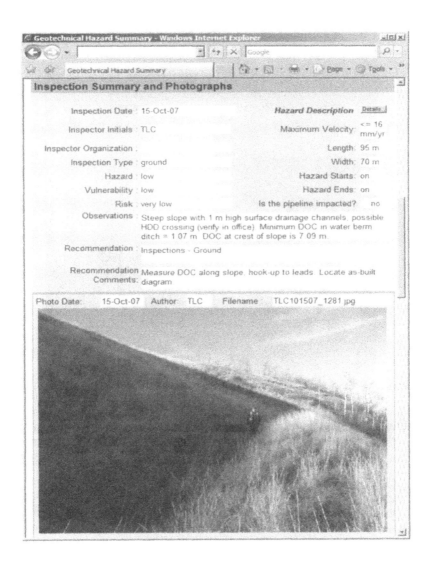

Figure 3 - Example of a field inspection form accessed from a geohazard data management system. (Published with permission from BGC Engineering Inc. as an update to figures presented in (Leir, 2004))

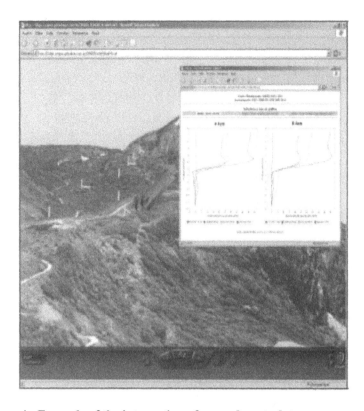

Figure 4 - Example of the integration of several route data sources and slope inclinometer results which appear when clicking on the instrument location (da Rocha, 2004).

Practical Considerations Relating to Data Availability

There are practical business context issues that impact a pipeline operator's ability to gather site specific and distributed route data. Figure 5 presents a simplified summary of the relationship over the life-cycle of a pipeline between business context, accumulation of route data (Rizkalla, 2007). Such practical matters later enter into a pipeline owner's consideration of when and which method of a range geohazard risk assessment and integrity management planning approaches. As will be discussed later in this chapter, quantitative or probabilistic geohazard assessment approaches that require a large amount of site-specific and general route data are better suited for operational stages following construction once as-built characterization and a sufficient base of operational data are available.

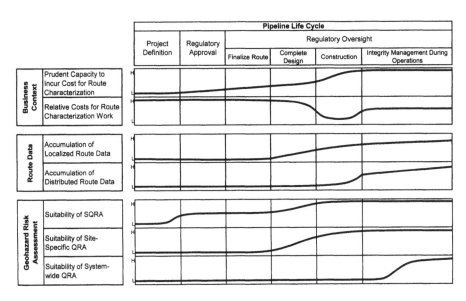

Figure 5 - The availability of route data through out the a pipeline life cycle and its impact on the methods of geohazard risk assessment and integrity management

6.4 Risk Assessment Methodologies

Risk Management in a Pipeline Context

According to Canadian Standards Association CSA Z662-03 (Annex B) risk management is the integrated process of risk assessment and control (CSA, 2003). Risk control is the process of decision-making for managing risk, and the related implementation, communication, and monitoring activities required to ensure the continuing effectiveness of the risk-management process. Risk assessment involves two elements: risk analysis and risk evaluation. Focusing on these two elements, the following specific definitions are provided:

- Risk analysis – the use of available information to estimate the risk, arising from hazards, to individuals or populations, property, or the environment. Risk analysis identifies what can go wrong, how likely it is, what are the consequences, and what is the level of risk.

- Risk evaluation – the process of judging the significance of the absolute or relative values of the estimated risk, including the identification and evaluation of options for managing risk. Risk evaluation determines if the risk is significant, and what options are available to manage the risk.

The risk analysis process involves hazard identification, frequency analysis, consequence analysis, and risk estimation. A hazard (or threat) in this context is defined as a condition with the potential to cause an undesired consequence. Methods for hazard identification fall broadly into three categories:

1) comparative methods such as checklists, hazard indices, and reviews of historical failures;

2) structured methods such as hazard and operability studies (HAZOP) and failure modes and effects analysis (FMEA); and

3) logic pathway methods, such as event tree or fault tree analysis, to link different release or initiating events to possible outcomes (CSA, 2003).

Frequency analysis is used to estimate the likelihood of occurrence of identified hazards, including associated impacts on the affected system. Results of the analysis are expressed in qualitative terms (e.g., failure very likely), or in quantitative terms either on a collective basis (e.g., failures per year) or on a linear basis (e.g., failures per kilometer year). Like hazard identification, several approaches for frequency analysis exist including:

1) Analysis of historical operational and incident data;

2) Fault and event tree analysis;

3) Mathematical modeling; and

4) Judgment of experienced and qualified engineering and operating personnel, based on known conditions (CSA, 2003).

The selection of a suitable analysis approach depends in part on the objectives of the assessment, and on the availability of applicable data and models.

Consequence analysis is used to estimate the severity of adverse effects on people, property and the environment resulting from such events as a release of toxic or flammable liquids and disruption of pipeline throughput. The effects of the release at any distance from the source can be estimated for the duration of exposure using mathematical models if the release mechanism and anticipated behavior of the released material are known. Results of the analysis can be expressed either qualitatively or quantitatively depending on the objectives of the assessment, data availability, and analysis approach adopted.

Risk estimation entails combining the results of frequency and consequence analysis to produce a measure of risk. Three approaches commonly used for risk estimation include:

1) risk matrix methods in which possible combinations of separate frequency and consequence (or likelihood and severity) estimates are presented in a two-

dimensional risk matrix of discrete risk categories – an example of a risk matrix is shown in Figure 6;

2) risk index methods in which factors that influence frequency and consequences are assigned values and mathematically combined to produce a risk index value; and

3) probabilistic risk analysis methods in which frequency and consequences are estimated quantitatively and mathematically combined.

		Consequences			
		Negligible	Low	Medium	High
Likelihood	Negligible	Negligible	Negligible	Very Low	Low
	Low	Negligible	Low	Low	Medium
	Medium	Very Low	Low	Medium	High
	High	Low	Medium	High	Very High

Figure 6 - Risk matrix example showing qualitative risk descriptions for combinations of likelihood and consequences.

Outputs from the matrix or index methods provide a relative measure of risk that is qualitative or semi-quantitative, whereas probabilistic methods produce absolute measures of risk that are expressed in quantitative terms such as cost/year. The selection of an appropriate risk estimation method depends on a number of factors such as stage of the pipeline life cycle, data availability, and level of detail required for the risk management decision-making process. All three categories are acceptable under Canadian Standard CSA Z662-03 if applied

255

in a rational and consistent manner. While Annex B of this standard indicates that risk assessment is not mandatory for pipelines, it is considered a valuable tool to support decision-making for pipeline integrity management. The outcome of such an assessment provides the pipeline operator with information required to decide where to intervene, when to intervene, and how to intervene to reduce the risk to pipeline integrity.

In summary, pipeline risk assessment involves identifying potential hazards to the pipeline, assessing their likelihood of occurrence and potential effects on pipeline integrity, and estimating the consequences associated with a loss of pipeline integrity resulting from the hazards or combination of hazards along the length of the pipeline. An additional consideration in assessing the potential effects of a hazard on pipeline integrity is the concept of vulnerability. (Fell et al. 2005) define vulnerability as the degree of loss to a given element or set of elements within the area affected by a hazard. Vulnerability of a pipeline element, for example, is an expression of the element's capacity to withstand a realized threat (i.e., hazard occurrence). This term is taken into account in order to determine whether or not a particular hazard occurrence is of sufficient magnitude to affect pipeline element integrity. Vulnerability ranges from 0 (no damage to the pipeline element in the case of a realized threat) to 1 (total loss of integrity of the pipeline element in the case of a realized threat).

Expressed quantitatively, the general expression for risk is as follows:

$$R = H \times V \times C \qquad\qquad \text{(Eq. 6-1)}$$

where H is annual hazard occurrence frequency or probability of hazard occurrence, V is vulnerability of the various system elements which determines the effect of the hazard occurrence on system integrity, C is the consequence associated with the hazard occurrence, and R is risk. In a fully quantitative risk assessment, risk can be expressed as an estimated annual cost associated with a particular hazard.

Geohazard Consideration in Pipeline Risk Management

Geohazards represent a special class of potential hazards or threats to a pipeline. Under American Society of Mechanical Engineers standard ASME B31.8S, geohazards are considered in a general manner under the category of Weather/Outside Forces. Canadian Standards Association CSA Z662-03 Annex Q identifies several loading mechanisms associated with geohazards. While these additional loadings are not specifically addressed in the Canadian standard, the designer is to determine if supplemental design criteria are necessary for such loadings, and if additional strength or protection against damage modes should be provided. According to the Canadian standard, the determination of geotechnical loads for design should distinguish between different types of load effects and load combinations, recognizing the current experience-based state-of-practice of geotechnical engineering.

In keeping with pipeline risk assessment approaches, geohazard risk assessment involves estimating the likelihood of occurrence of a damaging event related to a particular geohazard or combination of geohazards, and the potential damage or consequence of the event. The distinction between damaging and non-damaging events is related to the vulnerability (or resistance) of the pipeline element under consideration. For example, a landslide that occurs but does not cause a loss of pipe integrity is a non-damaging event in relation to the pipe. Such an event, however, may cause damage to the right-of-way that requires remediation, and hence is a damaging event with respect to the right-of-way. The consideration of pipeline element vulnerability is an important concept in identifying potentially significant differences between the probability of occurrence of a particular geohazard event, and the probability of failure of the pipe or other pipeline element in relation to the event. The role of vulnerability (or resistance) in differentiating between probability of damage (i.e., a hazard event reaching the pipeline) and probability of failure (i.e., loss of integrity as a result of the hazard event) is discussed further by (Muhlbauer 2006).

In the general geohazard literature, vulnerability tends to be considered as a scaling factor on consequences for particular system elements (e.g., Lee and Jones, 2004). In this approach, relations between the degree of loss of each system element and associated consequences are developed. As described by (Fell et al. 2005), this approach is well suited for localized geohazard phenomena, such as a landslide at a known location where the threat can be thoroughly characterized and analyzed, and the potential damage to system elements (i.e., buildings, utilities, vehicles, land, etc.) and any population in the area can be accurately estimated. However, the approach implicitly requires identification and characterization of each potential geohazard threat to the system elements at risk prior to consideration of system vulnerability. For long, linear, distributed systems in rugged terrain, identification and characterization of all potential geohazards regardless of possible impact on the system elements may be impractical, costly, and, in many cases, unnecessary.

An alternative approach for linear, distributed systems such as pipelines is to consider vulnerability in combination with the hazard frequency term in the risk equation. The product of hazard frequency and vulnerability is termed "susceptibility" in this approach. It is equivalent to likelihood of failure in qualitative or relative risk terms, or probability of failure (PoF) in quantitative or absolute risk terms. Zero susceptibility means that the pipeline element under consideration is not at risk (i.e., not susceptible to the imposed threat). Susceptibility can be zero for two reasons: absence of a threat (i.e., $H = 0$) or sufficient capacity of the pipeline element to withstand the imposed threat (i.e., V

257

= 0). By factoring the vulnerability of particular pipeline elements into an initial screening of pipeline geohazards, it is possible to screen out insignificant threats and prioritize those to which the pipeline element is susceptible. Of the pipeline elements, the right-of-way and ditch are generally more vulnerable than the pipe to some geohazards (e.g., erosion). To account for this difference in susceptibility and associated consequences, separate risk expressions are required to account for the risk to each of the pipeline elements (i.e., the pipe, ditch, and right-of-way).

The use of screening criteria based on the vulnerability concept allows differentiation of potentially damaging threats from benign ones in advance of detailed field characterization planning and deployment. Screening criteria based on pipeline element vulnerability may be related to a specific characteristic of a geohazard (e.g., the length of an unstable longitudinal slope or the amount of settlement expected due to thawing of permafrost) or to some combination of factors. By starting with a clear set of criteria related to specific geohazards, an initial screening of the pipeline route can not only identify potentially damaging geohazards, but can help prioritize field investigations and data collection activities. Emphasis can be placed on those geohazards that pose the most significant threat to the pipeline. This approach may also eliminate particular classes of geohazards if deemed non-threatening to the pipeline. For example, ice-wedge cracking in permafrost may not produce enough localized tensile strain to damage a robust pipeline design, and can therefore be eliminated from further consideration (NEB, 2006). The notion of a screening assessment in advance of a more rigorous hazard analysis is central to a staged risk management strategy (Eguchi et al., 2005). The decision to undertake detailed field characterization as part of pipeline risk management should be tempered with practical considerations such as the stage of the pipeline life cycle, and the cost/benefit of collecting detailed characterization data versus adopting conservative assumptions in the risk assessment.

Examples of vulnerability criteria for different geohazard threats are shown in Table 5. In each case, the critical value associated with the criterion is related in part to the pipe specifications. A number of closed-form solutions are available to estimate critical values associated with these thresholds (e.g., Sweeney et al., 2005). In other cases, soil-pipe interaction, hydrodynamic and other specific analyses are required to estimate critical values. Tools for such analyses include specialized software for stress and deformation analysis of cross country and offshore pipelines such as PIPLIN (PIPLIN, 1991), finite element software for detailed stress-strain analysis of soil-pipe-interaction such as ABAQUS (ABAQUS website), finite difference software for consideration of complex time-dependent behavior of soil such as FLAC (www.itascacg.com/flac), and other software to account for unique environmental conditions (e.g., geothermal analysis in arctic environments).

Table 5 - Examples of Vulnerability Criteria for On-shore Pipelines in Various Terrains

Geohazard Threat	Vulnerability Criteria	Basis	Application
Deep-seated longitudinal slope instability	Critical longitudinal slope length	Based on a proprietary closed form solution	Longitudinal slope stability assessment
Deep-seated transverse slope instability	Critical transverse slope width	Based on published closed-form solution (Sweeney et al., 2005)	Cross slope stability assessment
Vertical scour at river crossing	Maximum unsupported span at river	Based on Danish Hydraulic Institute software (PRCI, 2000)	River crossing analysis

PIPLIN is an example of an industry-recognized computer software program for stress and deformation analysis of above-ground and buried cross country and offshore pipelines. Several nonlinear aspects of pipeline behavior, including pipe yield, large displacement effects and nonlinear soil support, are accounted for by the program. Tabular output includes pipe displacement; anchor and soil support deformations and reactions; pipe axial forces, bending moments and curvatures; and axial and hoop stresses and strains in the pipe. Specific features of the software are as follows:

- The pipe is modeled with pipe elements. Stresses and strains are monitored at a number of points in the cross sections at the element ends. Pipe yield at the monitored points is taken into account assuming the von Mises yield criterion, with Mroz nonlinear kinematic hardening.
- Up to six additional coaxial layers may be assumed around the pipe. These materials may be brittle or ductile to simulate the behavior of concrete, insulation, watertight jackets, coaxial pipes or reinforcement.
- The soil is modeled as a nonlinear Winkler foundation. Discrete supports for anchors or above-ground bends are modeled as nonlinear springs.
- Pressure, temperature, settlement and gravity type loads are assumed to be applied statically. Arbitrary sequences of loads, including cyclic and non-proportional loads may be specified. The non-linear problem is solved by a step-by-step procedure with an event-to-event solution strategy. Equilibrium corrections are applied to compensate for nonlinear effects.

PIPLIN has been applied in a number of different applications including: side-bend and over-bend analysis of buried pipe with linear or non-linear soil restraint and rational treatment of anchorage length; settlement or excavation of trench bottom with arbitrary settlement profile and nonlinear soil including lift-off; two-dimensional fault motion of buried or above-ground pipe with lift-off and development of cable tension in pipe; analysis of the effects of combined longitudinal and transverse soil settlement caused by soil liquefaction or underground mining excavation; frost heave analysis of straight or bent pipe configurations; and computation of moment-curvature relationships for pipe with specified stress-strain curve including hoop stress effects due to internal pressure and axial stresses due to restraint of thermal expansion.

Non-linear soil springs are used in PIPLIN to represent soil loading and restraint conditions on pipelines exposed to large ground displacements. Standard equations related to pipeline-soil interaction are directionally-dependent, and are presented here for general information. Full treatment of the pipeline-soil interaction modeling subject can be found in several key references:

- "Guidelines for the Seismic Design of Oil and Gas Pipeline Systems" (ASCE 1984)
- "Guidelines for the Design of Buried Steel Pipe" (ASCE 2001; 2005)
- "Extended Model for Pipe Soil Interaction" (PRCI, 2003)
- "Seismic Design Guidelines" (PRCI,2004)

The reader is referred to these and other publications referenced in these documents for a more detailed description of the various analysis tools and their potential uses in risk assessment.

Axial Soil Springs

The maximum axial soil force per unit length of pipe that can be transmitted to the pipe is given by:

$$T_u = \pi Dac + \pi DH\bar{\gamma}\left(\frac{1+K}{2}\right)\tan(\delta) \qquad \text{(Eq. 6-2)}$$

where:

D = pipe outside diameter
c = soil cohesion representative of the soil backfill
H = depth to pipe centerline
$\bar{\gamma}$ = effective unit weight of soil

260

K =   effective coefficient of horizontal earth pressure which may vary from the value for at rest conditions for loose soil to values as high as 2 for dense dilative soils

α =   adhesion factor (Figure 7), a curve fit to plots of recommended values is

$$= 0.608 - 0.123c - \frac{0.274}{c^2 + 1} + \frac{0.695}{c^3 + 1} \quad \text{where c is in ksf or kPa/50}$$

δ =   interface angle of friction for pipe and soil $= f\phi_m$

ϕ_m =   maximum internal friction angle of the soil

f =   coating dependent factor relating the internal friction angle of the soil to the friction angle at the soil-pipe interface.

The corresponding displacement Δ_t at T_u for various soil conditions is given by:

Δ_t	=	3 mm for dense sand
	=	5 mm for loose sand
	=	8 mm for stiff clay
	=	10 mm for soft clay

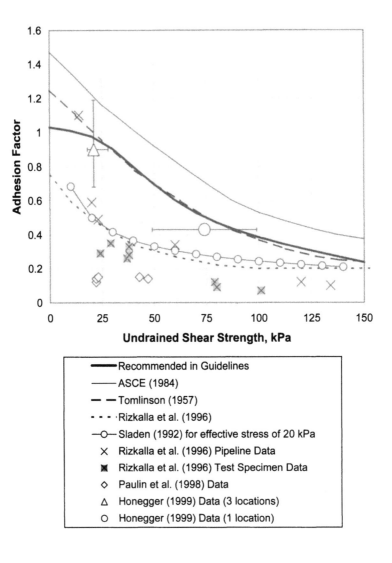

Figure 7 - Plotted values for adhesion factor, α showing the recommended relationship in American Lifelines Alliance (ALA) Guidelines (ALA, 2001)

Horizontal Soil Springs

The maximum horizontal soil force per unit length of pipe that can be transmitted to the pipe is given by:

$$P_u = N_{ch}cD + N_{qh}\overline{\gamma}HD \leq Q_d \qquad (6\text{-}3)$$

where:

N_{ch} = horizontal bearing capacity factor for clay (0 for $c = 0$)

$$= N_{ch}^* + 0.85\frac{\gamma H}{c} \leq 12$$

$$N_{ch}^* = 2.15 + 1.72\frac{H}{D} \leq 7.25$$

N_{qh} = horizontal bearing capacity factors for sand (0 for $\phi = 0°$)

$$= a + b\frac{H}{D}$$

c = soil cohesion representative of the soil backfill
D = pipe outside diameter
H = depth to pipe centerline
$\overline{\gamma}$ = effective unit weight of soil
Q_d = maximum vertical bearing soil force per unit length of pipe that can be transmitted to the pipe.

Values for a and b in the above relation for horizontal bearing capacity are summarized in Table 6. It should be noted that N_{qh} can be interpolated for intermediate values of ϕ between 35° and 45° and should not be taken as less than 35° even if soil tests indicate lower ϕ values.

Table 6 - Values for Horizontal Bearing Capacity Coefficients for Sand

ϕ	H/D Range	A	b	Maximum N_{qh}
35°	0.5 to 12	4	0.92	15
40°	0.5 to 6	5	1.43	23
	6 to 15	8	1.00	
45°	0.5 to 7	5	2.17	30
	7 to 15	10	1.33	

The corresponding displacement Δ_p at P_u is given by:

263

$$\Delta_p = 0.04\left(H + \frac{D}{2}\right) \le 0.10D \text{ to } 0.15D \qquad (6\text{-}4)$$

The foregoing relationships adopt an approximation to the findings for dry sand (Figure 8)

Vertical Uplift Soil Springs

The equations for determining upward vertical soil spring forces are based upon small-scale laboratory tests and theoretical models. For this reason, the applicability of the equations is limited to relatively shallow burial depths, as expressed as the ratio of the depth to pipe centerline to the pipe diameter (H/D). Conditions in which the H/D ratio is greater than the limit provided below require case-specific geotechnical guidance on the magnitude of soil spring force and the relative displacement necessary to develop this force. The maximum vertical uplift soil force per unit length of pipe that can be transmitted to the pipe is given by:

$$Q_u = N_{cv}cD + N_{qv}\bar{\gamma}HD \qquad (6\text{-}5)$$

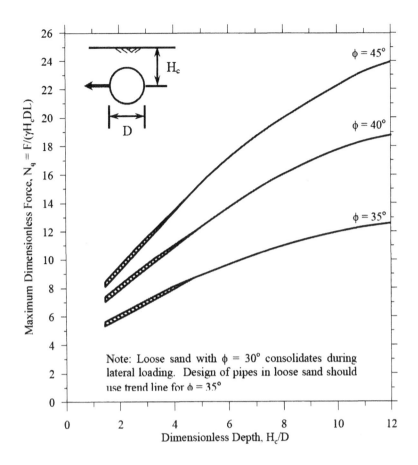

Figure 8 - Graphical presentation of values of N_{qh}

where:

N_{cv} = vertical uplift factor for clay (0 for $\phi = 0$)

$\quad = 2\left(\dfrac{H}{D}\right) \leq 10 \quad$ applicable for $\left(\dfrac{H}{D}\right) \leq 10$

N_{qv} = vertical uplift factor for sand (0 for $c = 0°$)

$\quad = \tan(0.9\phi)\left(\dfrac{H}{D}\right) \leq N_{qh}$

N_{qh} = horizontal bearing capacity factors for sand (0 for $\phi = 0°$)
c = soil cohesion representative of the soil backfill
D = pipe outside diameter

265

$$H \quad = \quad \text{depth to pipe centerline}$$
$$\overline{\gamma} \quad = \quad \text{effective unit weight of soil}$$

The corresponding displacement Δ_{qu} at Q_u is given by:

$$\Delta_{qu} \quad = \quad 0.01H \text{ to } 0.02H \text{ for dense to loose sands} \leq 0.1D$$
$$= \quad 0.1H \text{ to } 0.2H \text{ for stiff to soft clays} \leq 0.2D$$

Vertical Bearing Soil Springs

Vertical bearing soil springs are defined based upon the assumption that the pipeline behaves as a continuous strip footing. The maximum vertical bearing soil force per unit length of pipe that can be transmitted to the pipe is given by:

$$Q_d = N_c cD + N_q \overline{\gamma} HD + N_\gamma \gamma \frac{D^2}{2} \tag{6-6}$$

where:

N_c, N_q, N_γ = bearing capacity factors

$$N_c = \left[\cot(\phi + 0.001)\right] \left\{ \exp[\pi \tan(\phi + 0.001)] \tan^2\left(45 + \frac{\phi + 0.001}{2}\right) - 1 \right\} \tag{6-7}$$

$$N_q = e^{\pi \tan \phi} \tan^2\left(45 + \frac{\phi}{2}\right)$$

$$N_\gamma \quad = \quad e^{(0.18\phi - 2.5)}$$
$$c \quad = \quad \text{soil cohesion representative of the soil backfill}$$
$$D \quad = \quad \text{pipe outside diameter}$$
$$H \quad = \quad \text{depth to pipe centerline}$$
$$\overline{\gamma} \quad = \quad \text{effective unit weight of soil}$$
$$\gamma \quad = \quad \text{total unit weight of soil}$$

The corresponding displacement Δ_{qd} at Q_d is given by:

$$\Delta_{qd} \quad = \quad 0.1D \text{ for granular soils}$$
$$= \quad 0.2D \text{ for cohesive soils}$$

266

Geohazard Risk Assessment

As with general pipeline risk assessment, approaches to geohazard risk assessment can be qualitative, semi-quantitative, or quantitative in nature. A qualitative approach is one in which the geohazard frequency and consequences are expressed in descriptive terms to determine a qualitative description of relative risk (e.g., high, medium, or low). A semi-quantitative approach is one that involves qualitative categorization or index-based ranking to express the geohazard frequency, and a quantitative estimate of consequences, from which relative risk is determined. A quantitative approach is one that involves estimating an annual probability of failure associated with each geohazard, and combining the outcome with a quantitative estimate of consequences to estimate risk in terms of annual cost (i.e., loss of life, financial loss, or damage to property/environment).

The emphasis of pipeline geohazard risk assessment tends to be estimating susceptibility (i.e., likelihood or probability of failure) associated with various geohazard threats. The likelihood or probability of failure for each hazard, including geohazards, is estimated and combined to produce an aggregate estimate of the overall likelihood or probability of failure distribution along the pipeline. Consequence analysis is generally conducted separately under the premise that the most significant consequences are related to a loss of pipe integrity, and that the magnitude of the consequences are independent of the cause of the loss of pipe integrity. In general, this reasoning is acceptable for situations where pipe integrity is the primary driver for pipeline management decisions. However, in situations where geohazards represent the most significant threat category to a pipeline, the consequences to population, property and environment resulting from geohazards that adversely affect the ditch or right-of-way, but not pipe integrity, must also be considered. This secondary category of consequences is dependent on the specific geohazard, and is more akin to traditional geotechnical evaluation of landslide risk, for instance (Fell et al., 2005). Susceptibility and consequence assessment are considered separately in the following subsections.

Geohazard Susceptibility Assessment

Geohazard susceptibility assessment in the context of quantitative risk assessment (QRA) involves determining the probability of occurrence of specific geohazards, and evaluating their effects on pipeline elements, to estimate probability of failure associated with each element. Qualitative or semi-quantitative risk assessment methods seek to determine a relative measure of the likelihood of failure. Using the example of a landslide geohazard (e.g., Fell et al. 2005; Picarelli, 2005; Lee and Jones, 2004), there is a broad range of methods available to the geohazard specialist to determine likelihood or probability of failure, described as follows:

267

1. Historical Data - Review historical data within a well defined study area such a geomorphologic province, or areas with similar geologic characteristics, to assess the number and timing of previous events. This approach requires: a) ascertaining the extent to which environmental change during the period under consideration affects the assumption of uniform conditions over time, b) ensuring that only that portion of the historical record that is complete and accurate is used while trying to maximize the length of the usable record, and c) undertaking steps to overcome problems associated with small sample sizes producing erroneous estimates of probability (Lee and Jones, 2004). Where relevant historical data are lacking, this approach cannot be applied. Likewise, small sample populations of relevant data may preclude the use of this approach.

2. Triggering Events – Establish initiating thresholds between triggering parameters, such as rainfall or seismic activity, and the geohazard, then determine the frequency of these triggering thresholds through analysis of climatic or earthquake records to estimate geohazard probability (Lee and Jones, 2004). Thresholds can be determined through relatively simple correlation with known historical events, or through more sophisticated predictive modeling if the geohazard failure mechanism is understood and historic data are lacking. The probability of geohazard occurrence is the product of the annual probability of the trigger and the conditional probabilities of the subsequent geohazard response. (Lee and Jones 2004) note that some geohazards may be affected by weekly or monthly rainfall patterns, while others may be affected more by high intensity events of relatively short duration. Therefore it is important to ensure precipitation and other triggering data are subjected to appropriate analysis before incorporating results into a phenomenological model.

3. Empirical Correlations – Correlate outcomes of index-based ranking systems to observed phenomena to produce a predictive methodology. This generally involves developing simplified empirical relations that incorporate weighted parameters related to stability. Predictions of stability conditions based on these simplified relations can then be used as a surrogate for more complex stability analyses. Repeated analyses can be conducted to estimate geohazard frequency (Fell et al, 2005).

4. Expert Judgment – Use expert judgment involving experience, expertise and general principles to assign probabilities to scenarios in the context of a structured approach. Examples of structured approaches include event sequences consisting of initiating events, geohazard responses and outcomes, or more complex logic diagrams known as event trees. Expert judgment typically combines some quantitative evidence related to historic frequency of a particular geohazard with site-specific analysis to predict likely future events. Peer reviewed individual judgments, and multi-disciplinary

judgments, are preferred to judgments by a single practitioner working in isolation (Lee and Jones, 2004).

5. Probabilistic Simulation Models – Develop a probabilistic simulation model of a complex sequential geohazard process by a) establishing a mechanistic model b) assigning probability distributions to represent variability and uncertainty in the key parameters, c) develop a probabilistic prediction framework and simulation strategy such as static Monte Carlo simulations or dynamic time-stepping, and d) run repeated simulations to develop a stable frequency distribution (Lee and Jones, 2004). The frequency distribution is then used to calculate probability of failure.

6. Stability Analysis – Conduct repeated stability analyses using either combinations of deterministic parameters, or probabilistic parameter estimates, to determine the proportion of cases for which failure is predicted. An alternative approach is to estimate a reliability index based on an assumed statistical distribution. However this approach is sensitive to the assumed distribution (Lee and Jones, 2004). While probabilistic stability analysis for slopes is a well advanced topic area with an extensive body of published references (e.g., Nadim and Lacasse 2003, Nadim 2005), this approach requires a high level of competence with statistical analysis. It is suited to detailed analysis of individual hazard locations, but is less practical for use on long, linear, distributed systems.

When describing risk assessment methods, it is important to distinguish between quantitative risk results and the means by which probability of failure is estimated. The latter can use quantitative and/or qualitative information in a variety of ways, some very mathematically intense, to estimate a quantitative result. What makes a risk assessment quantitative is the form of the outcome, not the nature of the assessment method. The methods listed above can be used individually or in combination in conjunction with qualitative, semi-quantitative, and quantitative risk assessment approaches.

For hazard analysis, a distinction is made between catastrophic and progressive failure mechanisms. Catastrophic failure can be treated as a time-independent phenomenon with large deformation occurring "instantaneously", and is therefore amenable to prediction using limit equilibrium and other slope stability analysis techniques. In contrast, time-dependent deformation (creep) or progressive failure may complicate the prediction of hazard frequency or time-to-failure for a given slope. Creep deformation may occur gradually over a long period of time, possibly linked to other time-dependent or time-variable phenomena (e.g., rainfall). In addition, shear strength and other material properties may be time-dependent and either decrease or increase with time depending on in situ conditions. Mechanistic models used as the basis for prediction consequently should be compared against observed deformation and failure mechanics to ensure similar characteristics are evident in both predicted and observed responses.

Consequence Assessment

Consequence analysis provides estimates as to the severity of adverse effects, including one or a combination of the following: injury or loss of life, environmental impacts and property damage, repair costs, business interruption costs, fines, loss of reputation, and others. Methods for estimating consequence are improving rapidly and are well documented by (Muhlbauer 1996, 2004) and others (e.g. Pluss et al. 2000, Zimmerman et al. 2002, Sutherby et. al. 2000).

Prominent among the consequences associated with geohazard incidents are adverse environmental impacts. Such negative consequences may manifest in ruptures that occur in slopes near water crossings, in the water crossing themselves or in sensitive terrain.

A consequence modeling and assessment framework is a matter of corporate values and a reflection of the societal expectations of the political, regulatory and socioeconomic environments that a pipeline owner operates within. Generally, a corporate level consequences framework is adopted by a pipeline owner operator to which hazard managers, including those charged with managing geohazards, must adapt their respective hazard risk assessments to.

Quantitative Risk Assessment

An example of a quantitative risk assessment (QRA) is given by (Zhou et al. 2000). A methodology combining several of the above approaches was applied as part of a QRA of slope instability hazards in the pipeline integrity management context for an operating company. The methodology incorporated a rainfall-ground movement correlation for a series of creeping slopes instrumented with slope indicators in a remote area of Canada. The method accounts for both static variables and dynamic variables. It uses ground movement susceptibility index (GMSI) as a measure of relative susceptibility to ground movement based on weighted values for bedrock geology, surficial geology, slope angle, and groundwater to classify slopes. The methodology was applied to 1100 river crossing slopes to provide input into a spreadsheet-based pipe-soil interaction model (Trigg and Rizkalla, 1994). A probabilistic approach was used to calculate pipeline failure probability for each slope, hence the assessment is considered quantitative. The methodology is shown schematically in Figure 9. The parameters in Figure 4 are movement δ, rainfall in current month r_{mo}, rainfall in previous month r_{m-1}, rainfall two months prior r_{m-2}, and empirical coefficients c_1, c_2, c_3, and c_4 based on observed trends.

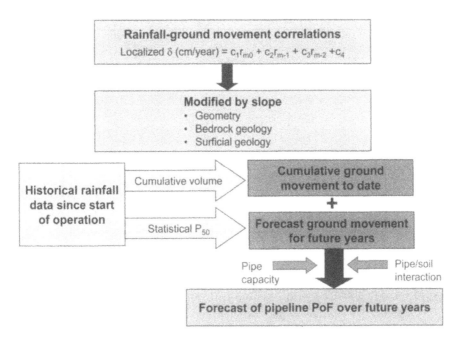

Figure 9 - Schematic of slope movement hazard analysis methodology applied by (Zhou et al. 2000a)

This example illustrates that, in order to account for the time variability of risk, one or more input parameters to the hazard analysis must be time-dependent or time-variable. Inputs such as rainfall correlated to movement are suitable for this purpose. This particular example demonstrated an increase in probability of failure with time calculated from the rainfall-slope movement correlation (Figure 10). This correlation allows prediction of critical slope movement

271

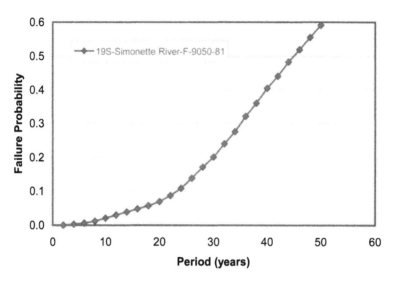

Figure 10 - Example of pipeline failure probability-time relationship based on predictive rainfall-slope movement model (Zhou et al. 2000a)

and planning of interventions as appropriate by introducing time variability into the analysis. Although not considered in this example, episodic variables such as earthquakes must also be considered in assessing hazard frequency.

Another example of quantitative geohazard assessment for pipelines is a methodology applied for pipeline integrity hazards due to hydrodynamic loadings at pipeline water crossings (Zhou 2000b). A Scour Hazard Database Model (SHDM) was developed to assess the risk of hydrotechnical threats to a pipeline with 2350 river and creek crossings. The approach considered both vertical and lateral scour. The lateral erosion analysis considered channel form, location, lateral cover distances between thalweg and pipeline, stream power, age of the crossing, and remedial work to develop a lateral scour rating value. Vertical scour analysis considered the age of the crossing, modeled scour, natural degradation and lateral pipe exposures at each crossing. Exposed pipes were evaluated based on the potential failure mechanisms including lateral instability, impact of debris, and fatigue to determine failure probability. A logic flowchart for vertical scour assessment is shown in Figure 11 as an example. Results of the analysis were used to produce quantitative estimates of failure probability due to hydrotechnical hazards as an input in proactive pipeline integrity management. .

272

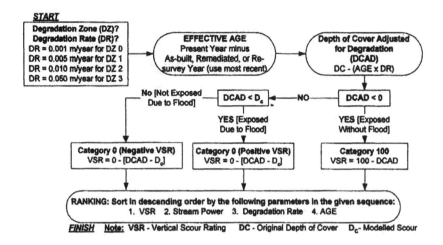

Figure 11 - Example of vertical scour logic flowchart used for hydrological effects analysis (Zhou et al. 2000b)

In the absence of correlations between observed or measured phenomena and hazard frequency, expert judgment can be applied to estimate a hazard frequency from qualitative classification data. (Muhlbauer 2006) recommends the use of logarithmic scales in estimating hazard frequency or failure probabilities to better reflect reality. This approach is equivalent to estimating probability to within an order of magnitude. A simple guidance chart is useful in attempting to assign a hazard frequency or failure probability (Table 7). Where enough data exist to develop correlations between failure frequency and qualitative or semi-quantitative, it is possible to map results from the qualitative or semi-quantitative assessment into a quantitative framework. For example, (Porter et al. 2006) propose a correlation between slope hazard class, qualitative priority and probability of failure for geohazard risk management of the Nor-Andino

Table 7 - Guidance Chart for Estimating Failure Frequency (after Muhlbauer, 2006)

Failures/year	Years to Fail	Approximate Rule of Thumb
1,000,000	0.000001	Continuous failures
100,000	0.00001	Fails ~ 10 times per hour
10,000	0.0001	Fails ~ 1 time per hour
1,000	0.001	Fails ~ 3 times per day
100	0.01	Fails ~ 2 times per week
10	0.1	Fails ~ 1 time per month
1	1	Fails ~ 1 time per year
0.1	10	Fails ~ 1 time per 10 years
0.01	100	Fails ~ 1 time per 100 years
0.001	1,000	Fails ~ 1 time per 1000 years
0.0001	10,000	Fails ~ 1 time per 10,000 years
0.00001	100,000	Fails ~ 1 time per 100,000 years
0.000001	1,000,000	Fails ~ 1 time per 1,000,000 years

Gas Pipeline. However, calibration and tests of reasonableness are essential if this strategy is adopted. The correlation between the qualititative/semi-quantitative likelihood estimates and quantitative probability of failure values will be project specific. The user is therefore cautioned not to universally adopt published correlations without conducting a test of reasonableness and ensuring there are enough calibration data to validate the correlation. If appropriately calibrated, such an approach can bridge the gap between qualitative, semi-quantitative and quantitative risk assessment.

Quantitative and probabilistic risk assessment approaches, where they may be effectively deployed, can present several potential advantages over other assessment methods. These advantages include (Fell 2005):

1. Consistency - imposing a rational and systematic approach to assessing safety and prioritizing integrity management activities through the use of well-defined assessment methodologies and analysis approaches to derive quantitative estimates of failure probability.

2. Compatibility - allowing comparison of risks across an owner's portfolio of hazards, such as geohazards versus metal loss processes. By expressing all hazards in a common absolute framework, the combination of co-spatial threats can be assessed directly from individual failure probabilities.

However, as was noted in Section 3 of this chapter on Data Management, the selection by an operator of a particular geohazard assessment approach is linked to practical business context issues that impact the operator's ability to gather site specific and distributed route data. Quantitative or probabilistic geohazard assessment approaches that require a large amount of site-specific and general route data are better suited for operational stages following construction once as-built characterization and a sufficient base of operational data are available. Even at this stage of development, however, there are practical limitations with respect to the level of information required to produce defensible estimates of failure probability. There are cautions in trying to implement a quantitative or probabilistic risk assessment of geohazards where insufficient data are available as input. As stated by (Lee and Charman 2004),

"The quality of a risk assessment is related to the extent to which the hazards are recognised, understood and explained – this is not necessarily related to the extent to which they are quantified. The temptation for increasing precision in the risk assessment process needs to be tempered by a degree of pragmatism that reflects the reality of the situation and the limitations of available information. Numbers expressed to many decimal places can provide a false impression of detailed consideration, accuracy and precision."

There remains a general lack of acceptance of quantitative and probabilistic risk assessment approaches within the geotechnical community except in specific cases where it is possible to gather sufficient data to conduct such analyses. As recognition of this state of practice, certain advanced pipeline design and risk management codes, such as those in Canada, acknowledge the practical challenges and limitations of applying these methods in practice. The Canadian code CSA Z662-03 allows for a range of assessment options including qualitative, semi-quantitative and quantitative.

Semi-Quantitative and Qualitative Assessment

Semi-quantitative risk assessment approaches seek to identify areas of potential susceptibility to specific hazards, including geohazards, and to rank the risk associated with these hazards in a relative manner. Such approaches typically involve categorizing parameter values or site conditions into discrete ranges to determine combined index values related to likelihood and consequences. These index values are then used to rank the relative risk associated with each identified threat. This approach is well-suited to linearly distributed systems where the level of data required for a defensible quantitative risk assessment may be insufficient to differentiate between natural variability in input parameters or ground conditions versus uncertainty in their estimation. In general, the level of information available in early project stages is considerably less than in later project stages where investigations have been undertaken to collect site-specific data, and as-built information is available, to supplement existing databases. Consequently, early project stages are often better suited to semi-quantitative or qualitative approaches to geohazard assessment whereby regional data can be used in conjunction with conservative assumptions about unknown conditions.

An example of a semi-quantitative risk assessment (SQRA) approach developed for pipeline systems on an industry consensus basis is provided by American Lifelines Alliance guidelines (Eguchi et al. 2005). A two phase process is recommended in which the susceptibility of the system to each threat is established before detailed, site-specific analysis is undertaken

- Phase 1 - Screening for susceptibility of the system to geohazards.
- Phase 2 - Detailed, site-specific evaluation of identified susceptible areas if warranted.

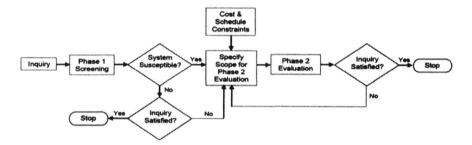

Figure 12 - Idealized flowchart showing the logic of assessing risks to linear systems such as pipelines (after Eguchi et al., 2005).

The scoring system used to develop a baseline level of analysis for performance assessments considers:

- Severity of the hazard (H)
- Vulnerability of the system or component (V)
- Damage consequences, including life safety (CLS), financial loss (CFL), disruption of service (CSD), and environmental and other impacts (CEI)

The level index (IL) is the product of individual severity indices for a specific threat, i.e., IL = H x V x max (CLS, CFL, CSD, CEI). Results of index level evaluation are used to select an appropriate level of investigation to address a specific inquiry. Investigation levels range from Level 1, which involves simple qualitative analysis, to Level 3 which involves more in-depth analytical approaches and quantitative analysis.

In keeping with the foregoing philosophy, and to account for the progressive accumulation of project specific data over the life of a pipeline, an evolutionary approach to geohazard assessment is considered prudent. By transitioning systematically from qualitative to semi-quantitative to quantitative assessment of geohazards over the project life-cycle, allocation of resources and expenditures between design, construction and operations can be balanced in relation to available route information. Early screening phases of the approach permit identification of areas requiring further investigation and possibly special site specific designs. Data collected in early stages of development are valuable inputs in planning field activities including locations and scope of field investigations. Early output from the geohazard assessment process also allows advance planning of possible monitoring/maintenance activities and frequencies during pipeline operations.

The general approach proposed for preliminary geohazard assessment of a pipeline involves the following tasks:

1. Framework development - develop an assessment framework including a list of credible probable geohazards, a susceptibility ranking methodology, and geohazard summary sheets describing the characteristics and identification/ranking criteria for each geohazard.

2. Spatial analysis - perform spatial analysis of geohazards using GIS-based queries for geohazard occurrence, assess susceptibility ranking of individual geohazards, and conduct dynamic segmentation of the route based on ranked geohazards

3. Susceptibility assessment (unmitigated) – produce a composite geohazard spatial distribution based on all credible probable geohazards, accounting for geohazard load combinations and trigger-event pairs (i.e., geohazards that reinforce one another), showing geohazard ranking for unmitigated conditions (i.e., prior to application of any mitigation options).

4. Mitigation strategy – identify options for reducing the rank of problematic geohazards through analysis, field characterization, design mitigation, and/or monitoring/maintenance, and estimate expected effects of applying the strategy.

5. Susceptibility assessment (mitigated) – produce a composite geohazard spatial distribution based on all credible probable geohazards, accounting for geohazard load combinations and trigger-event pairs (i.e., geohazards that reinforce one another), showing geohazard ranking for mitigated conditions (i.e., after application of the mitigation strategy).

Following this initial assessment, more detailed analysis can be undertaken at locations of high susceptibility.

Ranking Methodology

The general application of the geohazard assessment approach includes evaluating a number of semi-quantitative indices for each geohazard on an occurrence-by-occurrence basis along the pipeline route. The indices are calculated initially assuming an unmitigated pipeline design to determine where and what type of mitigation may be required. The expected effect of mitigation is then considered to determine a post-mitigation index value. This process links mitigation options (i.e., the toolbox) to in situ conditions and pipeline design, thereby illustrating the basis for selecting specific mitigation options (i.e., the decision tree). A flowchart illustrating the assessment process is shown in Figure 13.

The four indices that constitute the basis for determining susceptibility of the pipeline to different geohazards are evaluated using expert judgment, and are described as follows:

- Initiation Index (I_i) – this index ranges from 0 to 1. It characterizes the potential for the geohazard to initiate at a specific location along the route by comparing route conditions to a defined initiation threshold for that particular geohazard. A one-dimensional geohazard-specific occurrence map for each possible geohazard along the route is developed on the basis of this index.

- Frequency Index (F_i) – this index ranges from 0.001 to 1. It characterizes the potential number of occurrences of a particular geohazard at a specific location in relation to the life of the project. Observations of potential contributing factors are taken into account in determining an appropriate frequency index value for a particular geohazard.

- Rate Index (R_i) – this index ranges from 0.1 to 1. It characterizes the rate at which a particular geohazard and associated effects may occur, differentiating between rapid and gradual events/processes that may affect the pipe, ditch and/or right-of-way.

- Vulnerability Index (V_i) – this index ranges from 0.001 to 1. It characterizes the potential effects of a particular geohazard on the three main pipeline elements (i.e., pipe, ditch and right-of-way), differentiating between those geohazards that may potentially affect pipe integrity and those that may require either routine or non-routine intervention. The vulnerability criteria account for the design strain capacity and durability of the pipe, and will therefore vary for different pipeline designs.

The susceptibility rank (S) of the pipeline to an occurrence of a specific geohazard (G_i) at an interval along the route is calculated as the product of these four indices, i.e.,

$$S_{Gi} = I_i \times F_i \times R_i \times V_i \qquad (6\text{-}8)$$

The calculated susceptibility rank is a semi-quantitative measure of relative effects on the pipeline elements for a given geohazard. It should be noted that susceptibility may be insignificant if initiation of the geohazard is not possible (e.g., slope instability on flat terrain), if the geohazard may occur only rarely at a given location (i.e., occurrence requires a very low probability triggering event), or if the vulnerability of the pipe and other pipeline elements is negligible (i.e., a very robust pipeline design capable of withstanding all but extreme geohazards).

The possible index values are intentionally set orders of magnitude apart to help differentiate severe threats possibly requiring special design mitigation (e.g., slope stabilization measures) from those that can be addressed using practical pipeline construction practices (e.g., bedding and padding to eliminate possible indentation of the pipe by shallow bedrock). A susceptibility threshold of 0.001 is used for initial screening, although this value requires validation for different pipeline projects to ensure it relates to a condition requiring intervention or mitigation of some sort. Susceptibility values above the threshold are considered in the design process for possible mitigation.

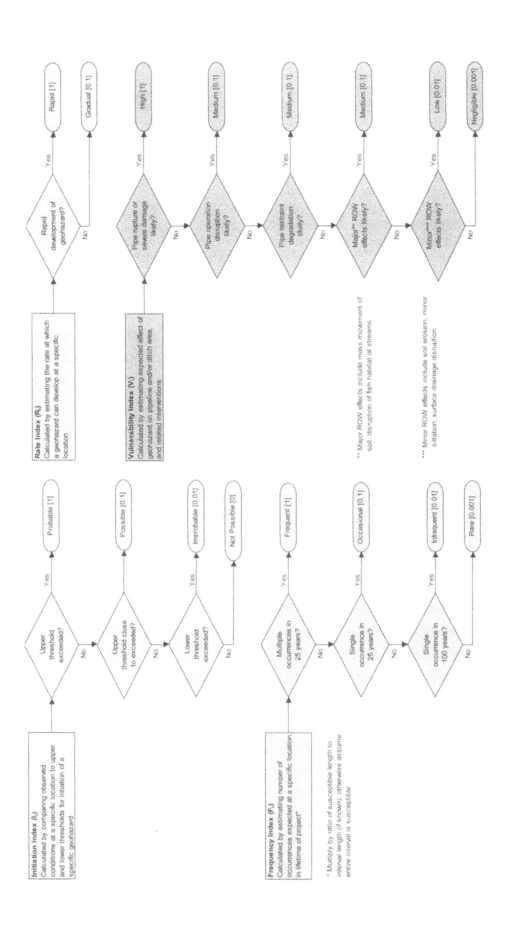

Figure 13 - Flowcharts for evaluating geohazard index values

Susceptibility Mapping

By systematically applying the susceptibility ranking methodology to the pipeline route for each credible probable geohazard, susceptibility maps and 1D takeoffs can be generated for individual geohazards. The next step in the geohazard assessment process involves producing a composite susceptibility map and 1D takeoff of the pipeline route on the basis of the individual susceptibility takeoffs. This composite takeoff identifies intervals that are either free of geohazards, contain a single geohazard, or contain multiple geohazards. The automated process of identifying intervals with unique geohazard characteristics is called dynamic segmentation of the route.

For those intervals containing multiple geohazards, further assessment is required to determine if the co-spatial geohazards represent load combinations (i.e., combinations that represent a more severe load on the pipeline elements than any of the individual geohazards acting alone), trigger-event pairs (i.e., combinations where one or more geohazards serve as a trigger for another more damaging geohazard), or simply coincident geohazards (i.e., geohazards that have unrelated effects on the pipeline elements). To conduct this assessment, a matrix of geohazard-pipeline interactions Table 8 was developed to categorize each geohazard in terms of four main effects:

- Pipeline deformation
- Pipeline boundary stress change
- Pipeline material/operational degradation
- Ditch and right-of-way effects

Table 8 - Example geohazard-pipeline interaction matrix

Effect of Geohazard on Pipeline Elements		Deep-seated landslide (longitudinal)	Deep-seated landslide (cross slope)	Slope creep	Rock fall/avalanche	Debris flow	Fault displacement	Dynamic liquefaction	Seismic ground motion	Vertical scour	Channel migration (lateral) (scour)	Buoyancy	Backfill erosion	ROW erosion	Karst collapse	Boulder/cobble/rock indentation
Pipeline Deformation	Axial Compression (shortening)	x														
	Axial Extension (lengthening)	x	x	x												
	Upward Displacement (flexure)											x	o		x	
	Downward Displacement (flexure)					x		x								
	Lateral Displacement (flexure)		x	x		x		x								
	Shear Displacement (shearing)	x	x	x			x									
	Point Loading (localized compression)															x
	Hydrostatic Loading (increased compression)															
	Hydrostatic Unloading (decreased compression)	x	x	x		x		x				x			x	
Pipeline Boundary Stress Change	Anisotropic Loading (increased compression)	x	x	x	x	x										
	Anisotropic Unloading (decreased compression)	x	x	x		x				x	x		o			
	Dynamic Loading (rapid oscillation)							x								
	Cyclic Loading (short- or long-period loading)								x							
	Hydraulic Loading (buoyancy)					x		x		x	x	x				
Material/Oper. Degradation	Corrosion (localized or general)	x	x	x		x		x		x	x	x	o		x	
	Exposure (loss of cover)	x	x	x		x		x		x	x	o	x	x		
Non-Pipe Effects	Ditch effect	x	x	x	o	x		x		x	x	o	x		o	
	ROW effect	x	x	x	o	x		x			x				o	

Legend: x - Direct effect o - Indirect effect

282

Table 9 – Example geohazard trigger-event pair matrix

	Primary Geohazard														
Secondary Geohazard ↓ / Primary →	Deep-seated landslide (longitudinal)	Deep-seated landslide (cross slope)	Slope creep	Debris flow	Rock fall/avalanche	Fault displacement	Dynamic liquefaction	Seismic ground motion	Vertical scour	Channel migration (lateral scour)	Buoyancy	Backfill erosion	ROW erosion	Karst collapse	Boulder/cobble rock indentation
Deep-seated landslide (longitudinal)	▨		t	t				t	t						
Deep-seated landslide (cross slope)		▨	t	t				t							
Slope creep			▨				t	t							
Debris flow				▨			t	t							
Rock fall/avalanche					▨			t							
Fault displacement						▨									
Dynamic liquefaction							▨	t							
Seismic ground motion						t		▨						t	
Vertical scour									▨						
Channel migration (lateral scour)										▨					
Buoyancy										t	▨	t			
Backfill erosion											t	▨			
ROW erosion													▨		
Karst collapse								t						▨	
Boulder/cobble rock indentation															▨

283

Comparison of pipeline effects for different geohazards is used as the basis for identifying possible load combinations and trigger-event pairs versus coincident geohazards. Matrices of geohazard trigger-event pairs Table 9 and of geohazard combinations were developed and are used to account for potentially higher susceptibility in intervals with multiple geohazards. The example matrices shown in Tables 9 and 10 cover a subset of the widely encountered geohazards; other less common geohazards can be added as required.

Although multi-branch chain-of-event effects are possible in terms of triggering mechanisms, the likelihood of geohazard occurrence diminishes with each additional branch in the conditional probability tree. The philosophy recommended as an assessment approach is to consider only two-step chain-of-event situations, and account for other possible contributing factors in assigning index values. The methodology to account for triggers involves reiterating the assessment based on the knowledge that one or more triggers are present at a specific location, and adjusting the index values as appropriate to recalculate susceptibility. To account for geohazard combinations, the susceptibility rank of combined geohazards in the interval is increased by one category (i.e., a factor of 10). In both cases, the susceptibility rank assigned to the interval is the maximum adjusted rank of all geohazards present within the segment.

The goal of the geohazard assessment is not to supplant a rigorous multi-disciplinary design approach, but to provide additional insight into spatial distribution and estimated severity of geohazards along the route. The output from the geohazard assessment is used in conjunction with other information as input to the design process.

Presented in Figure 14 is a stylized output of the assessment for several individual geohazards indicating a ranked distribution of their occurrences along a pipeline route. Based on these individual data bands, a composite data band may then be compiled by GIS queries integrating the total multi-hazard unmitigated susceptibility rankings distributed over the pipeline route. Subsequently, a second composite data band may be compiled reflecting the post-mitigation susceptibility rankings distributed over the pipeline route.

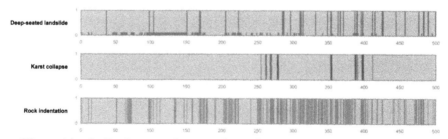

Figure 14 - Stylized output from geohazard assessment of pipeline route

284

6.5 Monitoring

The monitoring of ground movements and/or impacts on pipeline integrity is primarily a concern during the operations phase. Monitoring of the ground and the pipe is not commonly specified during design unless unstable ground cannot be avoided by routing modifications.

Where a geohazard mechanism is active on an operating pipeline ROW, the threat to pipeline integrity is the accumulation of strain in the pipe resulting from ground movements occurring around, below or above the pipe. Strains can increase to the critical limit states associated with high strain levels, ranging from deformation to pipeline rupture.

Pipeline owners seek to balance the following issues in managing ground movements:

- The need to minimize the consequences of pipeline failures, including public safety, public outrage, environmental impact, regulatory response and/or throughput interruptions or loss of revenue;
- The need to be cost effective in managing the hazard during operations including the costs to access remote areas or to work in harsh conditions;
- The practical challenges in implementing mitigative or remedial work when needed, ranging from scheduling an outage (flow interruption) to mobilizing the necessary personnel and equipment to isolated sites;
- The nature of the ground movement mechanism itself (ranging from episodic/seasonal small progressive movements to year-round small progressive movements to occasional large catastrophic movements) may dictate the nature and frequency of monitoring and may call for alternative construction methods;
- The need to accommodate land use and/or landowner related constraints.

In balancing these considerations, an operating pipeline company may elect to delay costly remedial measures, particularly if these measures require an outage and significant throughput impacts. Fortunately, in most cases the presence of ground movements does not create an immediate threat to pipeline integrity and ground movements usually do not require immediate stabilization unless site-specific conditions strongly indicate an impending slope failure or large movement. Instead, the goal is to determine the timing of remediation to maintain pipeline integrity, considering both the movement behavior of the ROW and the capacity of the pipe.

The presence of ground movements can be identified by their surface expression. Where the surface of a ROW is moving, this surface expression is commonly identified during aerial patrol. For example, a slope movement can be recognized by:

- tilting of trees or other vegetation in the downslope direction; and/or,
- surface tension cracking on the slope.

Once identified, an important next step in the assessment of the impact of ground movements on pipeline integrity is the development and implementation of a plan to monitor movements and pipeline impacts. Typically, the intent of monitoring a slope is to assess both the cause and effects of ground movements; i.e., to understand what triggered the movement and/or what natural processes are at work to maintain movements (cause); and, to assess the impact on the pipe (effect).

Given that the serviceability of the pipe is not immediately threatened, a monitoring plan is usually established to acquire either/both slope and pipe response data during operations. The objective of a monitoring program is to enable a pipeline company to intervene to mitigate the effects of a geohazard. Monitoring data is often solely relied on to make decisions on the timing and method of intervention. Where appropriate, this data is input to pipe-soil interaction modeling to determine the optimum time to intervene at a site.

As a complimentary reference source, the reader should review the discussion of geohazard monitoring technologies in the guidelines developed under the sponsorship of the PRCI and Department of Transportation. At the time of this book's publication, these guidelines were anticipated by 2009.

Monitoring ROW Geohazards

In deciding on the best method to monitor the progress of a geohazard, a pipeline engineer considers a range of proven technologies as summarized in Table 10 and described in subsequent sections of this chapter. Of particular interest is the connection between the available monitoring options in Table 10 and the individual geohazards summarized in Table 11 *i.e.*, which monitoring methods can be applied to the individual geohazards considered in this chapter. Table 10 represents a simple checklist that pipeline engineers can quickly reference to make a first determination of the likely monitoring methods to apply, given the specific geohazard that is being considered.

Aerial Methods

Aerial Patrol

Aerial patrol surveillance (APS) of the ROW is the most common method of detecting and monitoring ROW movements. APS identifies slope instability, erosion, channel scouring, bank slumping or other conditions needing remediation that are readily visible by low-altitude flights.

Maintenance personnel familiar with the long-term right-of-way and pipeline conditions participate in regular aerial patrols. The maintenance staff are trained to assess specific key indicators of ROW movements. Field observations are reported to engineering and environmental staff who assess the level of threat of each hazard identified in the aerial patrol and take appropriate action.

The follow-up to aerial patrol may include, depending on the threat level: site inspections; aerial photography interpretation of the site; a geological and geotechnical investigation program; and installation of instrumentation to track the progress of movements.

Table 10 - Overview of Geohazard Monitoring Technologies

Technology	Description	Merits	Limitations
Aerial Methods			
Aerial Patrol Monitoring	ROW surveillance technique ideally using dedicated observers in fixed-wing or rotary aircraft.	Common technique in the industry which generally has regulatory acceptance; frequency can be adjusted to conditions; cost effective for most pipeline systems; rotary aircraft can set down for surface examination.	Typically only identifies slope movements; no assessment of movement magnitude or depth; smaller movements may go undetected.
Air Photography Interpretation	Examination of pre- and post-movement air photography to identify movements and estimate movement rate	Technique is often used in geological and geotechnical evaluations; relatively inexpensive;	Only relatively large movements can be identified and assessed; photography may not be current or may not be available at all; requires specialized personnel;
LiDAR Survey	Laser-based system which provides dense topographic data from an aerial platform; repeat-pass comparison provides accurate measure of ground displacement	While relatively new, LiDAR is an established technology in many jurisdictions; accurate and repeatable;	Generally expensive; primarily measures vertical displacements; accurate estimate of slope movements is likely more difficult;
Satellite-based Methods			
Interferometry Using Radar Imagery (InSAR)	Use of repeat-pass radar imagery and interferometry to measure slope movements and subsidence;	Remote sensing technique requiring no field presence; particularly apt in remote locations where several sites could appear on one image; image acquisition frequency fits into creep-type movements.	Not applicable to full range of slope aspect azimuths; ground instrumentation may be required to mitigate effects of vegetation growth; image resolution may not be adequate for accurate estimate of movements;
Optical Imagery	Use of repeat-pass optical imagery to identify slope movements, river migration or subsidence	Remote sensing technique requiring no field presence; colour imagery is available.	Ground movement magnitudes or rates generally cannot be estimated; requires relatively large movements

288

Table 10 – Continued

Technology	Description	Merits	Limitations
Ground Instrumentation			
Slope Inclinometers	A custom PVC tube is installed in a borehole; electronic probe is raised through the tube by small increments; slope movements identified and monitored by measuring deformations of the tube.	An industry standard.	Requires drilling rig and crew; moderate environmental impact; generally requires field presence to monitor; prone to wildlife impacts. Requires replacement in areas of active ground movement.
SONDEX Tubes	The depth to a series of steel rings is measured by a peak voltmeter response. These rings move downward in response to settling ground. Depths are compared to a baseline series of soundings to determine vertical displacement.	A moderately simple system using conventional technology. The results indicate both the location and magnitude of settlement.	Requires a borehole to install an access pipe, resulting in a potentially expensive technique, particularly in remote areas.
Extensometers	Extensometers are used to measure settlement and heave in a wide range of facilities.	Comparatively accurate measurement of local deformation using tiltmeter probe; borehole drilling enables sampling of geological conditions; relatively common geotechnical instrumentation;	Requires drilling rig and crew; moderate environmental impact; generally requires field presence to monitor; prone to wildlife impacts.
Support Instrumentation	Primarily ground water/soil pore pressure instrumentation but should comprise any instrumentation that supports the analysis of the cause of the movement.	Provides input to slope stability analyses; common geotechnical instrumentation.	Requires drilling rig and crew; moderate environmental impact; generally requires field presence to monitor; prone to wildlife impacts.
Seismograph	Also known as a seismometers, a seismograph is a sensitive instrument that detects seismic waves generated by earthquakes, nuclear explosions or other seismic sources.	The most common monitoring device to detect the onset and characteristics of ground motions that may impact pipelines.	Warning of a significant seismic event may not be sufficient to prevent damage to an operating pipeline.

Table 10 – Continued

Technology	Description	Merits	Limitations
Conventional Survey			
Precise Ground Surveys	Establish a grid of pins or markers to monitor surface movement; locations of pins established by routine surveys;	Relatively low cost with low environmental impact; often used as the first monitoring method to confirm presence of movements.	Only provides movement on ROW surface; have to interpret impact of surface movements on pipe strain; pins/markers are subject to destruction by wildlife; requires attendance by field crew.
Underwater Diver Survey	Where appropriate the degree of vertical scour in a watercourse channel section may be assessed using divers to examine and photograph the crossing conditions.	Relatively low cost; low environmental impact; photographic record can be obtained.	Estimate of scour rate requires two or more divers' surveys; velocity of stream currents may be too great to safely use divers; visibility in a stream may be poor.

Table 11 - Geohazard Monitoring Summary

Geohazard	Geohazard Monitoring Methods			
	Aerial Methods	Satellite Methods	Ground Instrumentation	Conventional Survey
1. Landslides/Mass Movement				
Deep-seated landslide	✓	✓	✓	✓
Slope creep	✓	✓		✓
Slope creep rupture	✓	✓		
Thawed layer detachment [1]	✓			
Rockfall or rock avalanche [2]	✓			
Debris flow [3]	✓			
Snow avalanche [3]				
2. Tectonics/seismicity				
Fault displacement (fault crossings) [3]	✓	✓	✓	
Dynamic liquefaction [3]				
Dynamic ground motion [3]				
3. Hydrotechnics				
Vertical scour (water crossings) [4]				✓
Channel migration (water crossings)	✓	✓		✓
Buoyancy (pipe uplift)	✓	✓		
Rapid lake drainage [3]	✓	✓		
Coastal inundation and flooding	✓	✓		
4. Erosion				
Backfill erosion (upheaval disp.)	✓			
ROW erosion	✓			
Subsurface (piping) erosion	✓			
5. Geochemical				
Acid rock drainage [5]				
Karst collapse	✓	✓	✓	✓
Saline soil/bedrock [5]				
6. Freezing of unfrozen ground				
Frost heave (pipe) [9]				✓
Frost heave (ditch)				✓
Frost bulb development (cross-country) [10]				
Frost bulb development (water-crossing)				
Frost blister				
Ice-wedge cracking				
7. Thawing of permafrost terrain				
Thaw settlement (pipe)	✓	✓		
Ditch settlement	✓	✓		
Thaw settlement (ROW)	✓	✓		
Thaw bulb (slopes) [6]				
Thermokarsting/massive ice				
8. Unique Soil Structure				
Boulder/cobble/rock indentation [8]				
Static liquefaction [3]				

Table 11 – Continued

Geohazard	Geohazard Monitoring Methods			
	Aerial Methods	Satellite Methods	Ground Instrumentation	Conventional Survey
Sensitive and residual soil [3]				
9. Desert and Sand Sea Geotechnics				
Dune migration	✓	✓		
Flash flooding/scour at wadis [3]				
10. Volcanic Activity				
Ash falls [3]				
Lahars [3]				
Pyroclastic flow [3]				

1 Shallow failure – instrumentation not cost effective
2 Aerial Patrol only
3 Too rapid to monitor for prediction

4 Typically a underwater diver survey
5 Natural chemical hazard – no ground monitoring possible
6 Piezometer instrumentation

7 Natural chemical hazard – no ground monitoring possible
8 Pipe monitoring only
9 Heave rod on pipe; typically monitored by inertial ILI
10 Thermistor instrumentation

Air Photography Interpretation

Conventional air photography can be used to monitor ground movements. This technique is particularly effective in assessing slope movements, river bank migration and other geohazards where relatively large movements occur with a horizontal component.

LiDAR Survey

A typical LiDAR system rapidly transmits pulses of light that reflect off the terrain and other objects. The return pulse is converted from photons to electrical impulses and collected by a high-speed data recorder. Since the formula for the speed of light is well known, time intervals from transmission to collection are easily derived. Time intervals are then converted to distance based on positional information obtained from ground/aircraft GPS receivers and the on-board Inertial Measurement Unit (IMU) that constantly records the attitude (pitch, roll, and heading) of the aircraft.

LiDAR systems collect positional (x, y) and elevation (z) data at pre-defined intervals. The resulting LiDAR data is a very dense network of elevation postings. The accuracy of LiDAR data is a function of flying height, laser beam diameter (system dependent), the quality of the GPS/IMU data, and post-processing procedures. Accuracies of ±15cm (horizontally) and ±15cm (vertically) can be achieved.

Ground deformations are measured by comparing a current LiDAR survey with a baseline survey. An example of the use of LiDAR in monitoring pipelines is found in (Ager et al, 2004).

Satellite-based Methods

Interferometry Using Radar Imagery (InSAR)

Using a technique known as 'interferometry' ground movements can be measured to sub-centimeter accuracy by processing repeat-pass synthetic aperture radar (SAR) imagery. The combination of interferometry and SAR imagery has created the InSAR measurement technique (Rizkalla and Randall, 1999).

Since SAR image data contains information on both phase (ϕ) and magnitude (A) of the backscattered radiation, topographic information can be derived from the difference in phase between two images. InSAR is based on the combination of two complex (phase and magnitude) and co-registered (accurately aligned) radar images of the same area from an almost identical perspective. The phase difference for each pixel is a measure for the local difference of the relative change in distance between the scatterer (the ground surface) and the SAR antenna (i.e., changes in topography may be derived from phase differences between two SAR images).

Using the same beam mode, a SAR satellite, such as Canada's RADARSAT-1, returns to the same location every 24 days, resulting in an update to ground movements at this frequency. More frequent visits are possible but considering that the majority of slope movements experienced on pipeline rights of way are slow-moving, this frequency is considered adequate.

In temperate climates, where rainfall and vegetation growth may influence the accuracy of interferometric estimates of ground movement, radar reflectors may be deployed to produce an unambiguous movement interpretation. An example of a reflector is shown in Figure 15.

Examples of the application of InSAR to ground movement monitoring are found in (O'Neil and Samchek, 2002), (Morgan et al, 2004 and Froese, 2004).

Figure 15 - Radar Reflector for InSAR Ground Movement Monitoring

Optical Imagery

An optical image is equivalent to a colour photograph from space. Optical imagery is commercially available in resolutions less than 1 meter which enables identification of ground changes with repeat-pass image processing.

Where significant ground movement has occurred on a ROW it may be possible to identify these movements by comparing two optical images. Accurate estimates of movement magnitude may be difficult to determine. An example of optical imagery is provided in Figure 15.

Ground Instrumentation

Slope Inclinometers

Slope inclinometer monitoring is a common technique for monitoring the depth and magnitude of movements within a slope with relatively high accuracy. A custom plastic tube is installed in a near-vertical borehole that passes through an unstable slope and is anchored in stable strata at depth. During a movement survey, a sensitive probe using two biaxial servo-accelerometers, is drawn upward from the base of the casing to the top. The probe is halted at half-metre intervals to acquire tilt measurements. The tilt measurements are converted to horizontal movements and the data at each interval is compared to a baseline to identify movement zones and to calculate movements through time.

A schematic of a slope inclinometer installation is shown in Figure 16 and typical output from a slope indicator survey is shown in Figure 17. A case history describing the use of inclinometers for slope monitoring in a pipeline application is presented in (Lanziani et al, 2007).

Figure 15 - Example of Optical Imagery

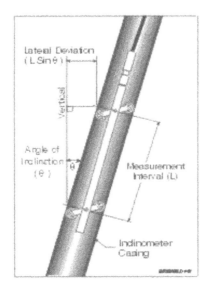

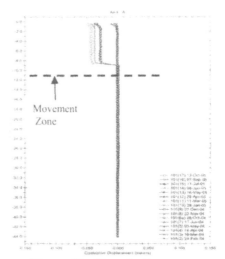

**Figure 16 - Slope Indicator
Installation**

**Figure 17 - Slope Indicator
Output**

SONDEX Tubes

The Sondex system measures vertical displacements. It consists of a probe, signal cable, a cable reel with a built-in voltmeter, and a number of stainless steel sensing rings that are positioned along the length of an access pipe. A survey tape is typically connected to the probe. In soft ground, sensing rings are fixed to a continuous length of corrugated plastic pipe which slips along the access pipe and allows the rings to move with the surrounding ground. In harder ground, rings can be attached directly to telescoping plastic pipe.

To obtain measurements, the operator draws the probe through the access pipe. The buzzer sounds when the probe is near a ring, and the voltmeter reading peaks when the probe is aligned with a ring. The operator refers to the survey tape and records the depth of the ring. A sensitivity adjustment allows operation adjacent to steel pipes, piles, or other metal objects. Settlement and heave are calculated by comparing the current depth of each ring to its initial depth.

A schematic of a Sondex tube installation is shown in Figure 18.

296

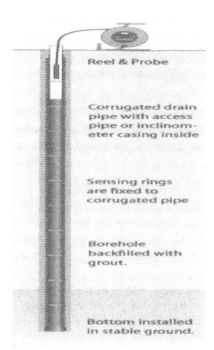

Reel & Probe

Corrugated drain pipe with access pipe or inclinometer casing inside

Sensing rings are fixed to corrugated pipe

Borehole backfilled with grout.

Bottom installed in stable ground.

Figure 19 - Rod Extensometer

Figure 18 - Sondex Tube Installation

297

Extensometers

Extensometers are used to monitor settlement or heave in excavations, tunnels, foundations, retaining walls and dams. The data from extensometer surveys indicate the depth of settlement and the total magnitude of settlement.

A number of different extensometer technologies are available. Rod extensometers are comprised of downhole anchors, rods with protective tubing and a reference head at surface. The anchors, with rods attached, are installed in a vertical borehole. The rods extend from the anchors to the reference head and the vertical position of the rod changes as the anchor settles. Settlement measurements are obtained with a depth sensor or depth micrometer. Settlement in a zone is calculated by comparing current positions of the rod with baseline measurements. A typical rod extensometer installation is shown in Figure 19.

Support Instrumentation

In addition to direct measurement of movements, instrumentation may also be installed to collect data for the numerical analysis of ground movements. For example, piezometers to measure pore pressures required in a slope stability analysis; or, thermistors to measure ground temperature for thermal analysis.

Conventional Survey Techniques

- **Precise Ground Survey**

  Monitoring of ground movements using survey techniques is common practice in pipeline operations. Typically metal pins are installed within and outside the suspected movement zone and movements are measured using electronic distance measuring (EDM) equipment. Conventional ground surveys can detect vertical as well as slope movements.

- **Underwater Diver Survey**

  The measurement of loss of cover due to vertical scour in a watercourse channel can be undertaken using a survey crew sounding the channel bottom from a boat. Alternatively, divers can be deployed in the channel to examine the pipeline alignment, particularly if exposure of the pipe is suspected. Divers can take accurate measurements of the exposure and photograph the surveyed area if required.

- **Other Techniques – Detection During Design**

  While not strictly methods of monitoring geohazards, it's worthy to note certain proven and common technologies used to detect the presence of geohazards during the design and operations phases.

- **Geophysical Methods**

  The application of geophysical methods is now well-established practice in the industry to investigate or detect the presence and extent of geohazards. For example

5.1 Monitoring the Pipe

The monitoring methods summarized in Table 12 have been applied to the categories of geohazards shown in Table 2 to a create summary table of geohazard monitoring. This summary is provided in Table 13.

- **Strain Gauge Instrumentation**

  Strain gauges are a direct-strain measurement technique available to assess pipeline strain. In this application, strain gauges are attached directly to the pipe during operations at a location where strain is known or suspected to be accumulating.

  Deployment of strain gauges during construction is not possible since the locations of potential high strain areas will not be known to the required degree of accuracy.

  Excavation of the pipeline is necessary to attach strain gauge rosettes. These are typically deployed at the 4, 8 and 12 o'clock positions on the pipeline. Depending on site specific conditions, four-gauge installations may be required.

  Figure 20 shows a strain gauge installation on a pipe. Discussions of the use of strain gauges for pipeline monitoring are presented in (Song et al, 2006 and Malpartida, 2007).

299

Figure 20 - Example Strain Gauge Installation

Table 12 - Summary of Proven Monitoring Technologies

Pipeline Monitoring

Technology	Description	Merits	Limitations
Strain Gauges	Gauges installed directly on the pipeline at $^1/_8$ or $^1/_4$ points around the circumference.	Provides direct measurement of strain; enables remote monitoring; supports pipeline structural modeling.	Requires pipe excavation; Gauges are prone to failure under field conditions. Provides spot strain measurements only. Difficult to ensure gauges are installed at the high or critical strain areas.
Inertial In-line Inspection (ILI)	An established pigging technology that measures pipe curvature changes and pipe strain accumulation.	Estimates pipe strain accumulation accurately; industry and regulatory acceptance; low environmental impact	ILI runs are relatively expensive for areas of few discrete geohazards. Requires pig launchers/receivers. ILI runs may be intrusive to throughput.
Fiber Optic Sensors	A fiber optic (FO) sensor is attached directly to the pipe. Changes in pipeline strain are recorded as changes in the FO response.	Accurate in location and magnitude of strain. Several successful implementations have been reported at the time of publishing this book and may be referenced in the proceedings of IPC 2008.	Requires pipe excavation. Depending on installation type or length, may only provide spot strain measurements. Difficult to ensure gauges are installed at the high or critical strain areas
Other Pipe Instrumentation	Includes custom-designed on-pipe instrumentation such as: heave or settlement rods; and, survey monuments for high-accuracy triangulation surveys. Instrumentation is isolated from the backfill.	Heave or settlement rods are relatively easy to install; survey measurement techniques represent proven technology; depending on the application, the pipe coating may not have to be removed.	Requires pipe to be excavated. Rods are mainly effective in measuring purely vertical movements.

301

Table 13 - Pipeline Monitoring Summary

Geohazard	Pipeline Monitoring Methods		
	Strain Gauges	Inertial ILI	Fibre Optic Sensors[i]
1. Landslides/Mass Movement			
Deep-seated landslide			✓
Slope creep	✓	✓	✓
Slope creep rupture		✓	✓
Thawed layer detachment		✓	✓
Rockfall or rock avalanche		✓	
Debris flow		✓	
Snow avalanche		✓	
2. Tectonics/seismicity			
Fault displacement (fault crossings)	✓		✓
Dynamic liquefaction	✓		✓
Dynamic ground motion	✓		✓
3. Hydrotechnics			
Vertical scour (water crossings)		✓	
Channel migration (water crossings)		✓	
Buoyancy (pipe uplift)	✓	✓	✓
Rapid lake drainage		✓	✓
Coastal inundation and flooding		✓	✓
4. Erosion			
Backfill erosion (upheaval disp.)	✓	✓	
ROW erosion	✓		
Subsurface (piping) erosion			
5. Geochemical			
Acid rock drainage		✓	
Karst collapse	✓	✓	
Saline soil/bedrock		✓	
6. Freezing of unfrozen ground			
Frost heave (pipe)		✓	✓
Frost heave (ditch)			
Frost bulb development (cross-country)			
Frost bulb development (water-crossing)			
Frost blister			
Ice-wedge cracking			

Table 13 - Continued

Geohazard	Pipeline Monitoring Methods		
	Strain Gauges	Inertial ILI	Fibre Optic Sensors[#]
7. Thawing of permafrost terrain			
Thaw settlement (pipe)		✓	✓
Ditch settlement			
Thaw settlement (ROW)			
Thaw bulb (slopes)		✓	
Thermokarsting/massive ice		✓	
8. Unique Soil Structure			
Boulder/cobble/rock indentation		✓	✓
Static liquefaction		✓	✓
Sensitive and residual soil			
9. Desert and Sand Sea Geotechnics			
Dune migration			
Flash flooding/scour at wadis			
10. Volcanic Activity			
Ash falls			
Lahars			
Pyroclastic flow			

Inertial In-Line Inspection (ILI)

Deformation monitoring ILI tools, also known as inertial tools, are designed to monitor the internal pipe geometry and location of its centerline (Hart et al, 2002). Inertial tools are typically equipped with the following instrumentation:

- Mechanical calipers to measure pipeline internal geometry (i.e., diameter, ovality, dents buckles and wrinkles, and weld identification;
- An Inertial Measurement Unit (IMU), which includes angle rate gyros and liner accelerometers, to measure the path of the tool through the pipeline;
- Odometers to measure the distance traveled (chainage) by the tool and instantaneous tool speed; and,
- Pressure and temperature sensors to measure pipeline operating conditions.

The objective of ILI using an inertial tool is to accurately estimate locations of maximum axial tensile and compressive strains resulting from ground movement loading. Axial pipeline strains have two components: uniform axial strain; and, flexural strain.

Uniform axial strain can be estimated by taking the measured weld-to-weld distance as determined by the odometer and calipers, subtracting the original joint length and dividing by the original joint length. To estimate flexural strain requires an assessment of the curvature of the pipe, which can be calculated by numerically differentiating the gyroscope and odometer data. Flexural strain is then calculated by subtracting the original pipe curvature from the ILI curvature estimate and dividing by the original curvature.

The industry's experience with the inertial tools demonstrates that ILI provides an accurate measure of pipeline deformation and through analysis axial strain in areas of significant ground movement. A graphic of a typical inertial tool is shown in Figure 21. The use of ILI in pipeline applications is described in several references including (Hart et al, 2002, Czyz, Wainselboin, 2003, and Czyz et al, 1996). A detailed description of the analysis of inertial ILI data to estimate bending strains is provided in (Czyz and Adams, 1994).

Figure 21 - Inertial Tool (Published with permission from BJ Pipeline Services)

A well presented example of the application of inertial ILI monitoring for the management of ground movement hazards for an NPS 30 pipeline was presented by Czyz and McClarty (Czyz and McClarty, 2004). The case study described the role of inertial ILI monitoring of a pipeline traversing a 420m wide creeping slope that led to the lateral deformation of the pipeline as shown in Figure 22 of the pipeline section during the subsequent strain relief excavation. The ILI monitoring program of that area spanned the period of 1994 to 2002. Samples of the data and subsequent strain analysis that may be derived from this monitoring technology are presented in Figures 23 and 24.

Figure 22 - The strain relief excavation of the NPS 30 pipeline after the pipeline deformation was identified by inertial ILI shows the extent of deformation   Source: (CZYZ and McClarty, 2004).

305

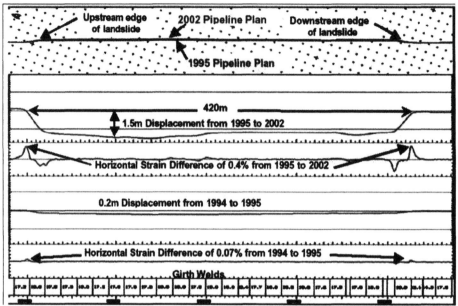

Figure 23 - ILI data and strain analysis results reported. Source: (Czyz and McClarty, 204)

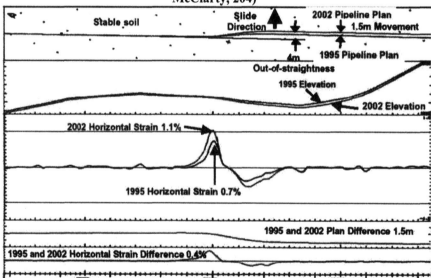

Figure 24 - ILI data and strain analysis results reported. Source: (Czyz and McClarty, 2004)

Fiber Optic Sensors

Pipeline strain monitoring by fiber optic has been the subject of development for some time (Rizkalla, 1993). More recent advances have brought the technology to a wide range of practical applications (Tennyson et al, 2004, and Chou et al, 2004). At the time of this book's publications , several cases of strain monitoring of buried pipelines were proposed for presentation in the 2008 International Pipeline Conference and associated papers were to be included in that conference's proceedings.

The fiber optic (FO) sensor is made from single-mode optical fiber, with a diameter of 250 microns. This small diameter allows the sensor to be bonded to a structure with minimum intrusion. The fundamental principle behind the FO sensor is low-coherence interferometery. The system can be considered as two optical paths combined with a light source and a detector. Each of the optical paths has a reflective mirror at the end, so that any light traveling down that path is reflected back. One path length is a reference length internal to the instrument, and the other path length is the FO sensor itself. When the system is at 'zero', with no load on the sensor, the two optical path lengths are exactly equal. When the sensor is under strain, the fibre is elongated (or compressed), resulting in a change in the optical path length. This change means that the two signals at the detector are no longer in phase, resulting in a drop in the magnitude of the signal. When this occurs, the system instrumentation changes the length of the reference path length until the two signals are in phase again. When the path lengths are matched again, the change in length of the sensor can be determined by calculating the change in length of the reference path. Since an interference peak only occurs within a very small distance (related to the wavelength of the light signal), the resolution of such a system is very high.

Other Pipe Instrumentation

In addition to the technologies described above, other pipe monitoring techniques include: heave or settlement rods attached directly to the pipe to monitor solely vertical displacements; and, high-precision triangulation surveys of on-pipe monuments to measure vertical and horizontal pipe movements.

The application of on-pipe instrumentation is typically limited to the operations phase. Not unlike the need of strain gauges to be installed at high strain location, the effective of on-pipe monuments is highly sensitive to being accurately located at critical section to the pipeline's structural response such as bends.

Planning of Monitoring Programs

There are usually unique considerations for each proposed ground or pipeline monitoring case. Generally, one or more of the following factors often influence the choice of monitoring approaches:

- Monitoring for cause and effect influences the type and number of instruments and installations;
- The accessibility or remoteness of the site: Is it accessible by ground, road, or strictly by air? Is the required support infrastructure nearby (hotels, vehicle rentals, fuel, etc.)? Will mobilization and demobilization costs be high?
- Will wildlife impact the on-going reliability of the installed devices? Does the ground surface conditions at and to the site enable easy access to instrumentation sites by wheeled or tracked equipment?
- Nature of ground movement and the spatial distribution of unstable areas of potential monitoring interest: Are additional unstable sites known or anticipated in the area? How rapid and large is the anticipated ground movement? How deep is investigation and monitoring required?

The monitoring plan will be influenced by the limitations of annual operating budgets and hazard management priorities set by the operating company.

6.6 Design and Operational Mitigation

Many of these hazards may be unavoidable during the initial route selection of a pipeline, while others could be aggravated by poor construction practices. In the case of pipeline routes traversing geologically-controlled, slope-instability-prone terrain or areas of permafrost or seismicity, avoidance of potential hazards to pipeline integrity may not be possible and special design mitigations and operational monitoring protocols will need to be developed.

As described earlier in the chapter, the results of an assessment of pipeline hazards is often a list of pipeline sections prioritized based on an analysis of risk. Those sections at the top of the list are often identified for the implementation of remedial or mitigative measures, to restore an acceptable risk level.

The objectives of mitigation planning are to:

- Minimize life-cycle costs, on a net present value basis
- Restore or maintain pipeline integrity
- Implement the mitigation in a safe manner
- Minimize throughput impacts.

When managing a geohazard in the design or operations phases, a pipeline company typically has two options to maintain pipeline integrity: reduce or eliminate the geohazard (the cause); or, reduce or eliminate the pipeline impacts (the effect). In some cases a combination of ground and pipeline mitigation may be the best solution, depending on the costs and risks of the available options. Generally, and with the exception of extremely sensitive environmental terrain, stabilizing the geohazard without proper assessment of the actual threat to the pipe should be avoided. Unless it's clear that the pipeline's integrity is in question, the prudent approach is often one that undertakes monitoring of the geohazard while undertaking even rudimentary analyses that account for the characteristics and condition of the pipe.

The decisions on when and how to mitigate a geohazard are often a result of a risk assessment of a pipeline system that accounts for all categories of hazards. These decisions typically balance the cost of a mitigation option with the estimated risk reduction resulting from a successful implementation. A discussion of the geohazard management planning process is presented in the following section of this chapter.

This section describes the options that may be considered when a geohazard has been encountered and a remedy is required; *i.e.*, the "tools in the toolbox" that have been generally proven effective in managing the geohazard itself or protecting the pipeline. To be useful to the practicing engineer, it's important to link each of the geohazards listed in Table 2 to one or more of the mitigation options in Table 14 to 17. As a complimentary reference source, the reader should also study the overview of geohazard mitigation methods presented in the guidelines developed under the sponsorship of the PRCI and the United States Department of Transportation. At the time of this book's publication, these guidelines were anticipated to be available in 2009.

Table 14 - Summary of Proven Mitigation Technologies - Geohazards

Technology	Description	Merits	Limitations
Ground Improvement			
Drainage	Drainage of a slope or soft ground is provided by shallow ditches, French drains or horizontally-drilled holes; ditch plugs to force water around the pipe to the surface; interception ditches installed at the top of a slope to prevent surface water ingress.	Drainage measures are cost-effective, proven technology.	Drilling of horizontal holes for drain installation is specialized. To be effective, a detailed sub-surface investigation is required to understand groundwater conditions so as to intercept the water bearing zones by the installed drainage measures.
Vegetation	Designed re-vegetation is applied to the RoW where erosion or shallow slope movements occur. Vegetation supports surface runoff without erosion and the roots contribute to the stabilization of shallow movements.	Proven technology with relatively low cost and a positive environmental impact. Specialized vegetation programs may be designed for particular slopes and soil conditions.	Should be restricted to erosion control and as a secondary line of defense against shallow instabilities.
Maintenance of Thermal Regime	In permafrost areas, specialized instrumentation (thermosyphons) are installed in frozen ground to maintain frozen conditions. Alternatively, surface treatment can maintain permafrost after the RoW vegetation is removed (e.g. wood chips). Foam insulation is applied above the pipe to maintain frozen ground	Proven technologies in permafrost areas. Passive systems requiring little maintenance. Options are suited for installation immediately after construction resulting in virtually no construction schedule impacts.	Thermosyphons require drilling or excavation equipment for installation. Wood chips from certain tree species are prone to self heating and should be avoided in RoW thermal protection applications.
Soil Densification	Where cohesionless soils (e.g. sands) exist in a loose state, liquefaction under ground shaking can occur; soils are densified by drainage, surface loading or vibrations to prevent liquefaction	Proven technology in site preparation of major civil developments.;	Likely expensive for narrow spatially distributed pipeline applications beyond for compressor or pump station sites.

310

Table 14 – Continued

Technology	Description	Merits	Limitations
Stabilizing Measures			
Slope Grading	To reduce movements the top of an unstable slope is lowered to reduce driving forces reducing or eliminating movements.	Involves simple earth moving– proven technology; reduces overbend angles on approach slopes	Graded material should be spoiled well back from the stabilized area.
Toe Berms	A berm is constructed at the toe of an unstable slope to increase the resistance to movement; attempts to slow stop movements.	Involves simple earth moving - proven technology;	Size and cross-section of of the berm requires special design. Land-use limitations or the presence of a water course may limit the cases where this solution may be applicable.
Retaining Walls	For small slope movements a retaining wall is installed at the toe of the movement to buttress the slope and reduce or stop movements	Proven technology; may be cost effective if local, native materials (e.g., rock) can be used;	A retaining wall is expensive in remote areas, particularly if native materials cannot be used; only effective in small movements;
Surface Diversion Berms	Small berms are constructed in orientations that divert surface water off the RoW; controls surface erosion	Relatively low cost if installed during pipeline construction; simple, proven technology	May require frequent maintenance if annual surface water flows are significant.
Channel Reinforcement	Where vertical or lateral scour is anticipated, the in-stream portion and/or the stream banks are armoured with rock, sandbags, vegetation, etc..	Proven, low-cost technology; and most likely less expensive than lowering an operating pipeline	In-stream reinforcement during initial construction or operations will require permitting and special measures to minimize stream and habitat contamination;
Re-establish Depth of Cover	Where a pipe is exposed (e.g., buoyancy, vertical scour or frost heave) but integrity is not threatened, the depth of cover is re-established in overland areas or in stream channels.	Proven technology; less expensive than re-burial of the pipe;	The logistics of completing this generally simple task in remote areas may lead to high costs. The design of the re-established cover has to consider the potential of future ot ongoing loss of cover.

311

Table 15 - Summary of Geohazard Mitigation – Geohazards

Geohazard	Design and Construction		Operations and Maintenance
	Ground Improvement	Stabilizing Measures	Stabilizing Measures
1. Landslides/Mass Movement			
Deep-seated landslide	✓	✓	✓
Slope creep	✓	✓	✓
Slope creep rupture			
Thawed layer detachment	✓	✓	✓
Rockfall or rock avalanche			
Debris flow	✓		
Snow avalanche			
2. Tectonics/seismicity			
Fault displacement (fault crossings)			
Dynamic liquefaction			
Dynamic ground motion			
3. Hydrotechnics			
Vertical scour (water crossings)			✓
Channel migration (water crossings)	✓	✓	
Buoyancy (pipe uplift)	✓		
Rapid lake drainage			
Coastal inundation and flooding	✓		✓
4. Erosion			
Backfill erosion (upheaval disp.)	✓		✓
ROW erosion	✓		✓
Subsurface (piping) erosion	✓		✓
5. Geochemical			
Acid rock drainage			
Karst collapse		✓	✓
Saline soil/bedrock			
6. Freezing of unfrozen ground			
Frost heave (pipe)	✓	✓	
Frost heave (ditch)	✓	✓	
Frost bulb development (cross-country)	✓		
Frost bulb development (water-crossing)	✓		
Frost blister	✓	✓	
Ice-wedge cracking	✓		

312

Table 15 – Continued

| Geohazard | Design and Construction | | Operations and Maintenance |
	Ground Improvement	Stabilizing Measures	Stabilizing Measures
7. Thawing of permafrost terrain			
Thaw settlement (pipe)	✓	✓	✓
Ditch settlement	✓	✓	✓
Thaw settlement (ROW)	✓	✓	✓
Thaw bulb (slopes)	✓		
Thermokarsting/massive ice		✓	
8. Unique Soil Structure			
Boulder/cobble/rock indentation			
Static liquefaction			
Sensitive and residual soil			
9. Desert and Sand Sea Geotechnics			
Dune migration		✓	
Flash flooding/scour at wadis			
10. Volcanic Activity			
Ash falls			
Lahars			
Pyroclastic flow			

Ground Improvement

Drainage

In geotechnical engineering, a primary consideration in stabilizing ground movements is the reduction in pore water pressures through a variety of ground drainage options, including:

- Shallow, open ditches where the unstable ground extends to depths less than 5 m;
- Ditches installed as French Drains, where the ditch is backfilled with drain material and perforated pipe to collect drainage water;
- Horizontal drains, installed with specialized drilling equipment, to release water from deeper unstable ground;
- In special cases involving large volumes of unstable ground, drainage using pumping wells or drainage galleries

In addition to dedicated drainage measures, the mitigation program may involve re-grading of the Right-of-Way (RoW) to promote surface runoff away from the RoW.

The geology of the unstable area will control the success of drainage systems. A comprehensive site investigation is normally required to understand the geology and hydrogeology of the area and to select the best drainage technique.

An example of successful drainage to stabilize pipeline Rights-of-Way is described in (Barlow, 2002).

Vegetation

Designed re-vegetation of the RoW by planting of a variety of species of grass and shrubs can improve the stability of shallow, unstable ground. Vegetation can be effective by several means:

- A thick mat promotes runoff, not infiltration, on sloping ground;
- Roots of the vegetation act as small 'anchors' in shallow soil; and,
- Certain vegetation has high water demand and can effectively drain unstable ground to root depth.
- The shaping of the ROW, including construction of runoff diversion berms, is typically undertaken in parallel with vegetation programs.

Maintenance of Thermal Regime

In permafrost areas, the prevention of permafrost thawing is important to prevent thaw settlement and/or shallow slope movements where thawed layers slide on permafrost. Maintaining the permafrost in a frozen condition can be accomplished by:

- Installation of ground cooling tubes, known as thermosyphons, is proven technology in northern regions to prevent thawing of frozen ground;
- Surface treatment that changes the thermal conditions at the ground surface has been proven on northern pipeline RoW(s) in maintaining permafrost; for example, wood chips from local trees have been effective in maintaining permafrost on thawing slopes; (Hanna et al, 1994);
- Placing foam board insulation above the pipe minimizes heat loss from the ground and minimizes thawing of the ground.

Soil Densification or Improvement

Densification of loose, granular soils is a technique used to prevent liquefaction, particularly during earthquake events. Though not common in pipeline applications, densification by vibrating needles, small explosives or by impact of large weights have proven effective in site development applications of less linearly distributed civil works. In addition to densification, improvement of loose soils by cement stabilization or other additives binds the soil and prevents liquefaction during shaking.

Stabilizing Measures

- ## Slope Grading

A common method to reduce movements on unstable slopes is to grade the ROW to a stable geometry. This process has the effect of reducing the ground force that is driving the movement. The objective of grading is to flatten the overall slopes and this can involve removing a portion of the top of the slope or grading to a flatter slope over the entire approach.

- ## Stabilizing Toe Berm

Another method of slope stabilization is the construction of a earthen berm at the toe of the slope. This improves the resistance to movement on the slope. Often slope grading and toe berm construction are co-incident, as the fill removed from the top of the slope is placed in the stabilizing berm at the toe.

- ## Retaining Walls

In the same way a toe berm increases the resistance to slope movement, so can the construction of a retaining wall at the toe of a movement. The caution is that retaining walls are effective only for small volumes of unstable ground. For intermediate to large movement areas, retaining walls would be required to be very large and expensive to resist the substantial earth forces acting at the toe.

- ## Surface Drainage Diversion Berms

Small earth berms are placed across the ROW in one of several patterns to channel or divert surface water off the ROW. Figure 29 shows an example of diversion berms placed across a sloping RoW

Figure 30 - Streambank Reinforcement

Figure 29 - Surface Water Diversion Berms

Channel Reinforcement

Where lateral channel migration has undermined the banks of a watercourse, bank repair should be properly engineered, as discussed in Chapter 3 of this book, and appropriate bank protection should be installed to prevent future scour. There is a wide range of channel bank protection options including:

- Properly-graded rock (rip rap) or rock baskets (e.g., gabions) are placed as a final lift along the bank slopes and extend into the watercourse; Figure 30 shows and example of a stream bank reinforced by gabion baskets.
- Locally available vegetation is used to reinforce existing vegetation on the bank slopes; for example, log walls comprised of local trees are erected along the bank as scour protection (Lees and Jalbert, 2006).

Re-establish Depth of Cover

Where the regulated depth of cover has been degraded, up to the point of pipeline exposure, depth of cover must be re-established to protect the pipe and maintain compliance with local codes. Depth of cover can be compromised in several ways:

- The pipe rises in saturated ground due to inadequate or lost buoyancy control;

316

- Frost heave, caused by freezing of thawed ground, causes the pipe lose cover;
- Vertical scour caused by flooding in a watercourse erodes the cover within a channel.

While in certain cases it may be logistically challenging to re-establish the depth of cover of a small section of pipeline in a remote area, the mitigation options themselves tend to be simple. Re-lowering-in the pipeline in its original ditch may be the more common and more preferred approach. However, in some cases, the option of adding material on top of the deformed pipe shape to meet the depth of cover requirements should also be considered.

Pipeline Mitigation

In a similar fashion to the section on mitigation of individual geohazards, this section describes techniques currently in use, and proven, in the pipeline industry for the mitigation of pipeline impacts alone. The "tools in the toolbox" for pipeline mitigation are summarized in Table 16. Note that the discussion of pipeline mitigation is independent of the type of geohazard that may have triggered the need for mitigation

As with geohazard mitigation, the link between individual geohazards and the categories of pipeline mitigation is important to narrowing the range of possible mitigation options practically available to a pipeline engineer. These links are summarized in Table 17.

Pipe Isolation

Deep Burial

This is a common mitigation method in pipeline design and operations where the pipe ditch is deepened in an area of shallow unstable ground. A deeper ditch enables the pipe to extend below shallow instabilities. In other cases, deeper burial of the pipe may provide additional cover to resist ground loading that is moving the pipe upward.

For example, deep ditch may be specified where a shallow, active layer movement on permafrost has been identified during route selection or may be required to resist frost heave forces where chilled pipelines encounter unfrozen, frost-susceptible soils. This category of mitigation also includes horizontal directional drilling (HDD). While this technique is typically used to mitigate environmental impacts at watercourse crossings, HDD may also provide a cost-effective means to provide deep burial in overland pipelines providing the HDD installation is assured to be below the unstable soil mass through a detailed site investigation

317

(possibly using inspection shafts to confirm the depth of the failure zone and the limit of the unstable mass). (Marcuz et al, 2007)

Table 16 - Summary of Proven Mitigation Options – Pipeline Protection

Technology	Description	Merits	Limitations
Pipe Isolation			
Deep Burial Horizontal Direction Drill	To mitigate effects of shallow slope movements and frost heave, etc. the pipe is buried to a depth that eliminates these effects.	Well-known solution that is suited for pre-determined localized areas.	Due caution is required in the field investigations and analysis to establish the appropriate depth of cover.
Synthetic Pipe Wrap	As a means to reduce the accumulation of ground movement induced loads a pipe wrap (sock) of geo-textiles is provided that allows the moving soil to slip along and not adhere to the pipe.	Cost effective and generally easy to install.	Only suited to cases where the primary direction of ground movement is longitudinal with respect to the pipeline
Straw Backfill	Use straw bedding and straw backfill in the ditch to provide a buffer of low-friction material between the pipe and the moving soil.	Typically a very low-cost solution and one that is easy to install.	Decomposition of material would be anticipated with repeated saturation cycles. Very difficult to consider in pipe-soil interaction modeling to assess longer-term performance.
Manufactured Backfill	Manufactured granular made from expanded shale and typically used as aggregate for light weight concrete has been used as a special frictional backfill material.	A significant reduction can be achieved in the magnitude of ground movement induced loads acting on pipelines in unstable terrain.	Likely to be a relatively costly option. Unless part of a wider trapezoidal cross-section ditch, this approach would be mainly suited to cases where the primary direction of ground movement is longitudinal with respect to the pipeline. In certain cases, questions have been raised about the long-term performance of the light weight aggregate material with respect to potential break-down to finer gradation.

319

Table 16 - Continued

Technology	Description	Merits	Limitations
Avoid Unstable Ground			
Pipeline Re-route	Pipeline is re-routed around the geohazard, through stable ground	Removes the pipe from the impact of ground movements.	Usually the most expensive mitigation option during operations. Very significant re-routes may be required to ensure that unstable ground is avoided.
Above-ground Pipelines	The pipe is raised above ground on supports in and beyond the zone of movement.	Removes the pipe from the impact of ground movements. Reduces or eliminates long-term remedial costs associated with unstable ground	Above-ground pipes are more susceptible to 3rd-party damage and incidents involving the public on the RoW. Should ensure the above-ground section extends for the full extent of the unstable zone before reestablishing the buried design mode.
Pipeline Ditch Modification			
Increased Ditch Width	A wider pipeline ditch is provided to increase the volume of compliant backfill surrounding the pipe and facilitate the pipeline's release from the ground (especially with trapezoidal ditch cross-sections)	Widely used in areas prone to seismic activity.	A slower ditching process is required. Not likely sufficient as the sole mitigative measure.
Bedding and Padding	Where the route encounters rock or boulders and cobbles in the ditch, special bedding and padding material is provided to prevent denting of the pipe during and after construction	A standard practice	Cost can be high if suitable bedding and padding materials is not found near the area of interest.
Pipe Excavation for Strain Relief	The pipeline section under strain from ground movements is excavated, allowing the accumulated elastic strain to be relieved.	A standard practice for relieving some of the accumulated pipeline strains	Costs are dependent on access and the possible need to reduce through-put during the excavation. Only the accumulated elastic strains are relieved.

Table 16 - Continued

Technology	Description	Merits	Limitations
Pipe Modification			
Heavy Wall Pipe	Increases the pipe's structural capacity and therefore increases the time before pipe mitigation is required;	A simple proven solution for some cases.	Typically very costly during operations
Pipe Bends - Vertical	Minimize the number and degree of pipe bends in unstable ground; for longitudinal slope movements, this reduces the points of compressive pipe strain.	A simple proven solution where practical (in terms of grading requirements)	May not be practical in certain cases.
Pipe Bends Horizontal	Minimize the number and degree of pipe bends in unstable ground; for lateral loading of the pipe, this reduces the points of pipe strain.	A simple proven solution where practical (in terms of grading requirements)	May not be practical in certain cases.
Pipe Installation Temperature	An increment of pipe strain is often the result of the difference in operating vs installation temperature, particularly in regions where construction is limited to cold, winter seasons. Warming the pipe during construction reduces the temperature difference and provides a greater strain allowance for ground movements.	Pipe is hoarded and warmed with conventional heaters and/or blowers; relatively simple operation.	Will certainly slow construction progress as pipe must be kept warm during welding, lowering-in and backfilling. Not suited for considerations during operations.
Cut outs	Where the pipe has experienced plastic strain (e.g. buckling) resulting from ground movement, the pipe section is cut out and replaced by new pipe.	A logical end-point in certain cases of operational maintenance. It should be noted that the pipeline's serviceability may be acceptable well beyond the elastic strain limit.	Costly, requiring a complete interruption of throughput (outage).

321

Table 17 - Summary of Geohazard Mitigation – Pipeline

Geohazard	Design and Construction			Operations and Maintenance		
	Pipeline Isolation	Route Alignment	Pipeline Ditch	Pipeline Isolation	Route Alignment	Pipeline Ditch
1. Landslides/Mass Movement						
Deep-seated landslide	✓	✓	✓	✓	✓	✓
Slope creep	✓	✓	✓	✓	✓	✓
Slope creep rupture						
Thawed layer detachment	✓	✓		✓	✓	✓
Rockfall or rock avalanche						
Debris flow						
Snow avalanche						
2. Tectonics/seismicity						
Fault displacement (fault crossings)				✓		✓
Dynamic liquefaction				✓		✓
Dynamic ground motion				✓		✓
3. Hydrotechnics						
Vertical scour (water crossings)					✓	
Channel migration (water crossings)	✓	✓				
Buoyancy (pipe uplift)	✓					
Rapid lake drainage						
Coastal inundation and flooding	✓				✓	
4. Erosion						
Backfill erosion (upheaval disp.)	✓				✓	
ROW erosion	✓				✓	
Subsurface (piping) erosion	✓				✓	
5. Geochemical						
Acid rock drainage						
Karst collapse		✓			✓	
Saline soil/bedrock						

Table 17 Continued

Geohazard	Design and Construction			Operations and Maintenance		
	Pipeline Isolation	Route Alignment	Pipeline Ditch	Pipeline Isolation	Route Alignment	Pipeline Ditch
6. Freezing of unfrozen ground						
Frost heave (pipe)	✓	✓		✓		✓
Frost heave (ditch)	✓	✓				
Frost bulb development (cross-country)	✓		✓			
Frost bulb development (water-crossing)	✓					
Frost blister	✓	✓				
Ice-wedge cracking	✓					
7. Thawing of permafrost terrain						
Thaw settlement (pipe)	✓	✓		✓	✓	✓
Ditch settlement	✓	✓			✓	
Thaw settlement (ROW)	✓	✓			✓	
Thaw bulb (slopes)	✓		✓			
Thermokarsting/massive ice		✓				
8. Unique Soil Structure						
Boulder/cobble/rock indentation						
Static liquefaction						
Sensitive and residual soil						
9. Desert and Sand Sea Geotechnics						
Dune migration		✓				
Flash flooding/scour at wadis						
10. Volcanic Activity						
Ash falls						
Lahars						
Pyroclastic flow						

Figure 31 - Synthetic Pipeline Wrap

Figure 32 - Straw as Pipeline Ditch Backfill

Synthetic Pipe Wrap

As demonstrated in a PRCI Study (PRCI, 2006), the use of two-layers of geotextile wrapping provides an effective means of reducing axial loads on buried pipelines subject to ground movements (see Figure 31). Wrapping the pipe with 2 geosynthetic layers (a woven geotextile in contact with the pipe wrapped by a non-woven filter fabric that will be in contact with backfill) decreases the frictional load due to soil movements by reducing the interface friction angle at the slippage surface. While the authors are aware of several field applications of this method, no publications were noted at the time of this book's publication reporting these field applications.

Straw Backfill

The use of straw as pipeline backfill on potentially unstable slopes has been utilized in several cases as shown in Figure 32. The potentially very low cost advantage of this pipe isolation methodology may be counter-balanced by the complexity of rigorously assessing its longer-term performance after the straw decomposition after several cycles of saturation.

Manufactured Backfill Materials

Use of special frictional material backfill has also been reported (Rizkalla, 1991) as shown in Figure 33. Commercially manufactured expanded (fired) shale products more typically used as aggregate for light weight concrete applications have been used as a special frictional backfill material. In some cases, questions were raised about the longer-term performance of the special backfill with respect to its potential breakdown to finer grained particles.

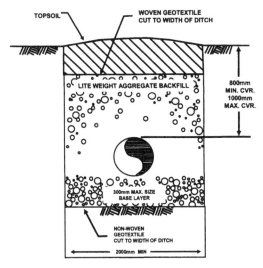

Figure 33 - Cross – section of ditch with special frictional material backfill

Avoid Unstable Ground

• **Re-route or Re-alignment**

Where unstable ground involves the entire RoW width, the pipeline can be re-routed around the unstable zone to avoid impact on the pipe. In other cases, an off-RoW movement may impact only a portion of the ROW width. To avoid these unstable areas may simply require the pipe to re-aligned to the stable side of the ROW.

Consideration of re-routing often involves a significant change and is generally an expensive alternative.

• **Above-ground Pipeline**

Where other ground movement or pipeline mitigation cannot be applied cost-effectively, the pipe may be brought above ground to bridge over the unstable zone. The above-ground section may be supported on wood or steel supports, or simply laid on the ground. An example of an above-ground pipeline is shown in Figure 34.

Pipeline Ditch Modifications

Increased Ditch Width

Widening the ditch through an unstable zone provides additional separation between the pipe and the moving soil outside the ditch walls. For a ground movement scenario in longitudinal directions, a wider ditch, especially when coupled with pipeline isolation measures, reduces the soil traction on the pipe. Similarly, where the pipe is being moved laterally, the force on a pipe may be somewhat softened by a wider ditch.

A trapezoidal ditch cross-section has been specified in cases where ground shaking due to seismic events may be expected. Such a cross-section, especially coupled with granular backfill, would be expected to allow the buried pipeline to be released from the surrounding ground during ground movements thus reducing the risk of potentially injurious loads being transferred to the pipeline.

Figure 34 - Above-ground Pipeline

Bedding and Padding

During route selection, the pipeline may be aligned through rock terrain or through surficial deposits that contain a substantial percentage of boulders and cobbles. When the ditch is excavated, the bottom and sides of the trench may contain hard and angular soils that can dent a pipeline during laying-in and backfilling. These dents may result in leaks during the commissioning or the operations phase.

326

Mitigation of this geohazard is provided by bedding, padding and or additional protective coatings. Bedding refers to natural or synthetic materials placed in the base of the trench prior to laying-in of the pipe to prevent contact with the pipe by boulders and/or cobbles. Bedding may be placed across the entire length of the trench base (e.g., with uniform sand) or may be established intermittently in the trench to 'bridge' over hazardous terrain (e.g., Styrofoam 'pillows'). Padding is the specification of finer grained materials as backfill in the ditch up to the pipeline's crown. A bedding and padding operation is shown in Figure 35. Additionally, where suitable bedding and padding materials are not economically available in a given area, special protective plant- or field-applied protective coating material may be specified.

Figure 35 - Placement of Bedding and Padding

Figure 36 - Pipeline Strain Relief

Pipe Excavation for Strain Relief

For pipelines subjected to accumulating ground movement leading to associated pipeline strain, a common mitigation technique is to excavate the pipeline thereby relieving the elastic pipeline strain. A sufficient length of pipeline should be excavated to allow for strain relief. Typically, the excavation of the pipeline allows it to "spring" from its initial position. For simple liner alignments, the extent of the strain relief excavation should be optimized in the field until no additional pipeline "spring" is observed. More detailed considerations are required in determining the extent of the strain relief excavation in cases of pipeline alignments that include significant bends.

327

The pipeline shown in Figure 36 has been excavated in this fashion and has deflected to the right side of the trench. (Sorensen, 2000) describes a case history where pipeline excavation and strain relief was chosen to mitigate the development of pipeline strain in a moving slope

6.7 Geohazard Management Planning Process

This section of the chapter is intended to bring together the various elements of geohazard management activities already introduced into an integrated process to develop actionable plans. A process flow is developed to link and balance between several activities towards establishing an annual geohazard integrity management program addressing known risks and monitoring results.

A shifting balance between inputs to geohazard management decisions

An interesting and somewhat complex process might ultimately unfold in the annual development of geohazard management programs of a long operating pipeline. At a higher level, there are 3 major sources of information that guide mitigation intervention decision over the lifecycle of a pipeline exposed to geohazards. These inter-related sources of information are:

Outputs of risk assessment analyses;

Field observations from reconnaissance and patrols by operating staff and specialists

Deformation and strain measurements from pipeline and ground conditions monitoring programs.

While these sources of information to integrity management decisions always remain as requirement to program development, the shifting balance in their respective prominence from the design and early operations phases of a pipeline project to later years of operations is indicated in Figure 37.

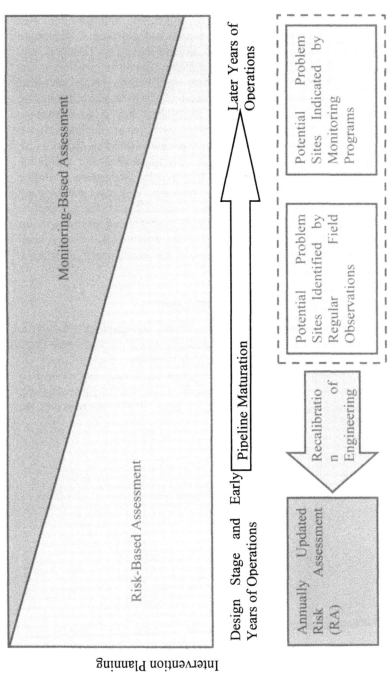

Figure 37 - Pipeline project to later years of operations

During a pipeline's design stage, with typically limited field information as was noted in Figure 5, an initial risk assessment should be undertaken. The design stage risk assessment guides where special designs are to be specified and identifies locations where monitoring instrumentation should be installed at the end of construction. With the start of operations, regular aerial patrols and other reconnaissance, both by operating staff and specialists, begin to provide inputs to the hazard management decision process. A balance should be established between more frequent aerial patrols by operational staff that are familiar with the right of way conditions and a prudent number of reconnaissances by geotechnical specialists to focus on special areas. In time, as operational decisions lead to the installation of site-specific monitoring, a wealth of pipeline and ground condition monitoring data is accumulated and ideally managed in a data management environment as was noted in section 3 of this chapter. Since the quantitative nature of the site-specific monitoring data is best suited to serve as the basis for more precisely timed mitigation interventions, there would tend to be a greater emphasis placed on that source of input data for the development of detailed integrity management plans.

Developing a geohazard management plan

How an integrity management program may actually be developed based on integrating the sources of information mentioned above will certainly vary from pipeline operator to another. However, for the purposes of a discussion to highlight several practical areas of interest, a generalized integrated process is suggested as presented in Figure 38 Eight areas of the flow process diagram, three of which addressing more than a single element of the diagram, will be discussed in turn.

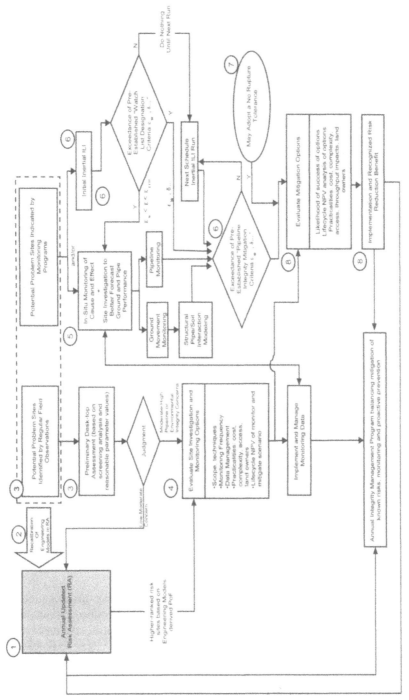

Figure 38 – Generalized Integrated Process for Developing a Geohazard Management Plan

331

(1) The range of potential geohazards has been presented in section 2 of this chapter while the approaches available to undertake geohazard risk assessments have been presented in section 4. A practical consideration to remain mindful of is the fit of the geohazard risk assessment process within the larger multi-hazard risk assessment process undertaken by a pipeline operator. As prudent pipeline operators will wish to develop a coordinated integrity management plan to address high priority risks across geohazards, metal loss, mechanical damage and other classes of hazards, it will become necessary to compare the risk rankings across different hazard classes. Due to the differences in the nature of the underlying physical models used to estimate probabilities of failure as well as difference in the state of practice across engineering disciplines, establishing a consistent basis for probability of failure comparisons might in some cases prove problematic. Further, depending on the calculation logic of multi-hazard risk assessment programs, the relatively localized risk exposure of some geohazards (e.g. slope instability) which may impact a 200m or 300m long slope may be difficult to compare to much longer manifestations of metal loss hazard mechanisms which are considered to impact a 20 km long valve section. Intrinsic difference between how engineering disciplines model different classes of hazards should be given due consideration.

(2) Considering geohazard risk assessment onto itself, the idealized or theoretical engineering model based algorithms for estimating probabilities of failure, should be reviewed and updated as required based on both the observations from aerial patrols and specialized reconnaissance as well as the lessons learned from monitoring programs.

(3) The decisions to react to potential geohazard problem sites identified through aerial patrol and specialized reconnaissance by undertaking site investigations and installing monitoring instrumentation should be guided by an engineering assessment based on screening analysis and reasonable parameter values. With respect to potential geohazard induced pipeline integrity issues, integrity management should be evaluated from the pipeline out – i.e. fully considering pipeline capacity as opposed to undertaking mitigation studies from a purely geotechnical focus. Prudent pipeline operators that adopt such a systematic approach remain must mindful of potentially adverse environmental impacts beyond pipeline integrity recognizing differences in solutions and timing considerations for both issues.

(4) When determined to be necessary, the evaluation of site investigation and monitoring options may be served in part by the overview discussion of monitoring presented in section 5 of this chapter and the aforementioned PRCI United States Department of Transportation guidelines. The available site investigation methods and their costs could vary widely according to the site-specific details of the area of interest. In fact, in certain cases where difficult seasonal access and other field work execution complexities may be present and

lead to very high costs, it may be prudent to compare the net present value of an option of first undertaking a site investigation including the installation of monitoring followed by an assumed mitigation intervention to the option of installing a conservatively designed mitigation together with monitoring without a site investigation.

(5) For sites established to be associated with a risk, monitoring of pipeline and ground performance is essential to understand both the cause and effect in play so as to better forecast when mitigative interventions will be required. Optimizations of monitoring frequencies, timely analysis as well as the effective management of collected data are all marks of prudent operators. Section 3 of this chapter provides an overview of data management tools to support this step.

(6) In cases where a pipeline operator's system is well equipped with in-line inspection (ILI) launchers and receivers or in cases where geohazards are sufficiently pervasive to warrant the installation of these facilities to support ILI inspections, the deployment if inertial ILI would typically be a great asset in managing several geohazards. Some of the interesting considerations in integrating inertial ILI in geohazard management are:

    a.   Establishing a baseline pipeline bending profile as soon as practical after construction

    b.   Determining the ILI run frequencies and adjusting as may be required based on the derived results

    c.   Working in a multi-discipline team environment, establish a strain criteria at which a localized section of a pipeline may be placed on a "watch list" which would then trigger a detailed site investigation and installation of monitoring equipment so as to gain an understanding of the cause and effect of the deformation and forecast a trend for future strain accumulation.

    d.   Also working within a multi-discipline team environment, establish a strain criteria, beyond the watch list criteria, which would trigger planning for and undertaking mitigative interventions, mindful of both the time to install the mitigation and its probability of success in arresting the underlying geohazard either immediately or over-time.

(7) In certain cases, prudent operating practice and high standards of corporate citizenship may support pipeline companies adopting a "No Rupture Tolerance". In such cases a higher ranking may be artificially assigned to an investment in risk mitigation than might have been otherwise arrived at through the risk assessment process.

(8) Typically, the design and implementation of mitigative interventions to address geohazard risks are major engineering and construction undertakings with significant associated costs. Section 6 of this chapter and the aforementioned PRCI United States Department of Transportation guidelines provide an overview

discussion of the range of mitigative options for geohazards. An essential step in the overall geohazard management process is updating the risk assessment database to reflect the risk reduction benefits for the area where the mitigative measure was installed. An equally important counter-balancing consideration however is to recognize the cases where the installed mitigative measure was partially or fully unsuccessful and to reflect the residual risk as not fully mitigated.

Section 6-8 References and Related Readings:

<http://www.**abaqus**.com> .

Ager, G., Marsh, S., Hobbs, P., Chiles, R., Haynes, M., Thurston, N., and Pride, R, "Towards Automated Monitoring of Ground Instability Along Pipelines." Int. Conf. Terrain and Geohazard Challenges Facing Onshore Oil and Gas Pipelines. Thomas Telford, London.

American Society of Civil Engineers (ASCE), "Guidelines for the Seismic Design of Oil and Gas Pipeline Systems, Technical Council on Lifeline Earthquake Engineering", Committee on Gas and Liquid Fuel Lifelines, New York, 1984.

American Society of Civil Engineers (ASCE), "Guidelines for the Design of Buried Steel Pipe", – report prepared for the American Lifeline Alliance (ALA), www.AmericanLifelinesAlliance.org. American Society of Civil Engineers, USA (with Annexes to 2005), 2001.

American Society of Mechanical Engineers (ASME) Managing System Integrity of Gas Pipelines: ASME B31.8S. USA, 2002. 66p.

Barlow, J.P., "Influence of Gradually Moving Slopes on Pipelines.", Proceedings of the International Pipeline Conference, Calgary, Alberta, Canada. Paper No. IPC 2002-27348, 2002.

Brunsden, D., "Geohazards and pipeline engineering," keynote paper, International Conference on Terrain and Geohazard Challenges Facing Onshore Oil and Gas Pipelines: Institution of Civil Engineers, London, UK, 2005.

Bruschi, R. et al.. Failure Modes for Pipelines in Landslides Areas. Pipeline Technology – OMAE, Canada, Volume V, pp. 65-78, 1995.

Bruschi, R. et al.. Pipelines Subject to Slow Landslides Movements Structural Modeling vs., Field Measurement. Pipeline Technology – OMAE, Canada, Volume V, pp. 343-353, 1996.

Canadian Standards Association (CSA)., CSA Z662-03 Oil and Gas Pipeline Systems including Annex B Guidelines for Risk Assessment of Pipelines, 2003.

Cappelletto, A., Tagliaferri, R., Giurlani, G., Andrei, G., Furlani, G., and Scarpelli, G., Field full scale tests on longitudinal pipeline-soil interaction. Proceedings of International Pipeline Conference 1998, Calgary, vol. 2, pp.771-778, 1998.

Charman, J.H., Fookes, P.G., Hengesh, J.V., Lee, E.M., Pollos-Pirallo, S., Shilston, D.T., and Sweeney, M., "Terrain, ground conditions and geohazards: evaluation and implications for pipelines," International Conference on Terrain and Geohazard Challenges Facing Onshore Oil and Gas Pipelines: Institution of Civil Engineers, London, UK, 2005.

Chou, Z.L., Cheng, J.J.R., and Zhou, J., "Monitoring and Prediction of Pipe Wrinkling Using Distributed Strain Sensors". Proceedings of the International Pipeline Conference, Calgary, Alberta, Canada. Paper 10595, 2006.

CONCAWE,. Western European cross-country oil pipelines 30 years performance statistics. Bruxelas, 2002. Relatório 1/02. Disponível em: <http://www.concawe.be>. Acesso em: 05 de março de 2005.

Czyz J.A., Fraccaroli C., Sergeant A.P., 1996, "Measuring Pipeline Movement in Geotechnically Unstable Areas Using An Inertial Geometry Pipeline Inspection Pig", 1st International Pipeline Conference, Calgary, Alberta, Canada, June 1996.

Czyz, J.A.; MacClarty, E.. Prevention of pipeline failures in geotechnically unstable areas by monitoring with inertial and caliper in-line inspection.. BJ Pipeline Inspection Services Technical Overview CD.

Czyz, J.A. and Wainselboin, S.E., "Monitoring Pipeline Movement and its Effect on Pipe Integrity Using Inertial/Caliper In-line Inspection", Proceedings of the IBP Rio Pipeline Conference and Exposition, Rio de Janerio, 2003.

Czyz, J.A. and Adams, J.R., "Computation of Pipeline-bending Strains Based on Geopig Measurements", Proceedings of the Pipeline Pigging and Integrity Monitoring Conference. Houston, 1994.

Czyz, J.A.; Wainselboin, S.E. Monitoring pipeline movement and its effect on pipe integrity using inertial/caliper in-line inspection. In: "RIO Pipeline Conference and Expostion, 2002, Rio de Janeiro. Anais... Rio de Janeiro: IBP, 2003.

da Rocha, R. S. et. al., "Geotechnical Data Management System of Transpetro Pipe Routes – Georisco", Proceedings of IPC 2004 International Pipeline Conference, October 4-8, 2004, Calgary Alberta, Canada.

Eguchi, R.T. et al., "Guidelines for Assessing the Performance of Oil and Natural Gas Pipeline Systems in Natural Hazard and Human Threat Events" – report prepared for the American Lifeline Alliance (ALA), http://www.AmericanLifelinesAlliance.org, 2005.

Esford, F., Porter, M. Savigny, K.W., Muhlbaure, K. A., Risk Assessment Model for Pipelines Exposed to Geohazards. In: 5th International Pipeline Conference, 2004, Calgary, Alberta, Canada, Proceeding: CD, 2004.

Federal Geographic Data Committee, Geospatial Positioning Accuracy Standards Part 3: National Standard for Spatial Data Accuracy, http://www.fgdc.gov/standards/projects/FGDC-standards-projects/accurcay/part3/index_html, 1998.

Fell, R., K.K.S. Ho, S. Lacasse, E. Leroi., "A framework for landslide risk assessment and management", Proceedings of the International Conference on Landslide Risk Management, Vancouver, Canada, 31 May – 3 June 2005, edited by Hungr et al., 2005.

Forese, C.R., Koooji, M.V.D., Kosar, K., Advances in the application of inSAR to complex, slowly moving landslides in dry and vegetated terrain. In: Ninth International Symposium on Landslides, 2004, Rio de Janeiro.: A.A. Balkema Publishers, 2004. 1746 p. p. 1255-1263.

Greaves T., Personal communication on the application of geohazard data management systems in Pembina Pipeline, 2007.

Hanna, A.J., J.M. Oswell, E.C. McRoberts, J. Smith, and T. Fidel. "Initial Performance of Insulated Permafrost Slopes, Norman Wells Pipeline

Project, Canada", In proceedings of the 7th International Cold Regions Conference, Edmonton, Canada, 1994.

Hart J.D., Powell G.H., Hackney D., Zulfiquar N, "Geometry Monitoring of the Trans-Alaska Pipeline", 11th International Conference on Cold Region Engineering, Anchorage, 2002.

Hengesh, J.V., Angell, M., Lettis, W.R. and Bachhuber, J.L., "A Systematic Approach for Mitigating Geohazards in Pipeline Design and Construction." Proceedings of the International Pipeline Conference, Calgary, Alberta, Canada, 2004.

Hungr O., Fell R., Couture R. and Eberhardt E. (eds.), Landside Risk Management. Proceedings of the International Conference on Landslide Risk Management, Vancouver, Canada, 31 May – 3 June 2005. A.A. Balkema, 2005.

<http://www.itascacg.com/**flac**> .

Lanziani, J.L., Massucco, G., and Valdivia, J., "Toccate's Slope, Monitoring and Control Activities" Proceedings of the Rio Pipeline Conference and Exposition, Paper No. IBP1355, 2007.

Lee, E. M. and D. K. C. Jones., "Landslide Risk Management" Thomas Telford Publishing, London, 2004.

Lees, A. and Jalbert, A., "Assessment of Successful Biostabilization Techniques on Selected Watercrossings in Alberta", Proceedings of the International Pipeline Conference. Calgary, Alberta, Canada, 2006.

Lee, E.M., and Charman, J.H., "Geohazards and risk assessment for pipeline route selection. Terrain, ground conditions and geohazards: evaluation and implications for pipelines," International Conference on Terrain and Geohazard Challenges Facing Onshore Oil and Gas Pipelines: Institution of Civil Engineers, London, UK, 2005.

Leir, M. and Reed, M., "Natural Hazard Database Application – A tool for Pipeline Decision Makers", Proceedings of IPC'02 4th International Pipeline Conference, September29-October 3, 2002, Calgary Alberta, Canada.

Leir, M., et. al., "Field Inspection Module for Hydrotechnical Hazards", Proceedings of IPC 2004 International Pipeline Conference, October 4-8, 2004, Calgary Alberta, Canada.

Lukas, A., Loneragan, S., and MacDonald, D., "The practicality of drilling very long pipelines under hazardous terrain—5 km, 10 km?" International Conference on Terrain and Geohazard Challenges Facing Onshore Oil and Gas Pipelines: Institution of Civil Engineers, London, UK, 2005.

MacCardle, A. et al.. Pipeline monitoring with interferometry in non-arid region. Rio Pipeline Conference & Exposition, 2005, Rio de Janeiro. Proceedings. Brazil: CD, 2005.

Malpartida Moya, J. E., 2007, Stress Monitroing in Pipelines Which are Susceptible to Soil Movements, Rio Pipeline Conference & Exposition, 2007, Rio de Janeiro. Proceedings. Brazil.

Manning, J., Willis, M., Denniss, A., and Insley, M. "Remote sensing for terrain evaluation and pipeline engineering," International Conference on Terrain and Geohazard Challenges Facing Onshore Oil and Gas Pipelines: Institution of Civil Engineers, London, UK, 2005.

Marcuz, G., Savigny, K.W. and Porter, M.J., "Landslide Avoidance Using HDD to Manage Social, Political and Technical Risks", Proceedings of the Rio Pipeline Conference and Exposition, Paper No. IBP1048, 2007.

Morgan, V., Kenny, S., Power, D., and Gailing, R., "Monitoring and analysis of the effects of ground movement on pipeline integrity," International Conference on Terrain and Geohazard Challenges Facing Onshore Oil and Gas Pipelines: Institution of Civil Engineers, London, UK, 2005.

Muhlbauer, W.K.. Pipeline Risk Management Manual. 3 ed. USA: Gulf Professional Publishing, 2004. 392 p.

Muhlbauer, W. K., "Pipeline Risk Management Manual, 4th Edition", Gulf Publishing Co., 2006.

Nadim, F., Einstein, H., Roberds, W., , "Probabilistic stability analysis for individual slopes in soil and rock, in Hungr", O., Fell, R., Couture, R., and Bernhard, E., eds., Landslide Risk Management, Proceedings of the 2005 International Conference on Landslide Risk Management: New York, A.A. Balkema, p.63-98. 2005.

Nadim, F. and S. Lacasse., "Probabilistic methods for quantification and mapping of geohazards", Proceedings of the 3rd Canadian Conference of Geotechnique and Natural Hazards, Edmonton, Alberta, Canada, 2003.

NEB Website, http://www.NEB.gc.ca/Application for the Construction and Operation of the Mackenzie Gas Project – GH-1-2004/Applicants/IORVAL/IORVL-145—Mackenzie Gas Project – Geohazard Assessment Workshop Materials, 2006.O'Rourke 2007.

Nadim, F., and Lacasse, S., "Mapping of landslide hazard and risk along the pipeline route," Terrain, ground conditions and geohazards: evaluation and implications for pipelines. International Conference on Terrain and Geohazard Challenges Facing Onshore Oil and Gas Pipelines: Institution of Civil Engineers, London, UK, 2005.

Oliveria, H.R., Vasconcellos, C.R.A., Freitas, J.D., A Historical Case in the Bolivia-Brazil Natural Gas Pipeline: Slope on the Curriola River. In: 5th International Pipeline Conference, 2004, Calgary, Alberta, Canada. Proceeding: CD, 2004.

Oliveira, H.R., A Proposed Approach for the Management of Pipeline Geohazard Risks, Masters of Engineering Dissertation (Translated to English) Submitted to the Federal University of Santa Catarina , Brazil, December 2005.

O'Neil, G., Samcheck, A., Satellite-based monitoring of slope movements on TransCanada's Pipeline System, Proceedings International Pipeline Conference, Calgary, Alberta, Canada, 2002.

Paulin, M.J., Phillips, R., and Boivin, R., Centrifuge modelling of lateral pipeline/soil interaction - Phase II. International Conference on Offshore Mechanics & Arctic Engineering, Copenhagen, June 1995, vol. V, pp 107-123.

Paulin, M.J., Phillips, R., and Boivin, R., An experimental investigation into lateral pipeline interaction. International Conference on Offshore Mechanics & Arctic Engineering, Florence, June 1996, vol. V, pp 313-323.

Paulin, M.J., Phillips, R., Clark, J.I., Hurley, S., and Trigg, A., Establishment of a full-scale pipeline/soil interaction test facility and results lateral and axial investigations in sand. Proceedings, 16th International Conference on

Offshore and Arctic Engineering (ASME - OMAE 1997), vol. 5, pp.139-246.

Paulin, M.J., Phillips, R., Clark, J.I., and Boivin, R., An experimental investigation into lateral pipeline/soil interaction - phase II. Centrifuge 98 - Proceedings of the International Conference Centrifuge 98, A.A. Balkema, Rotterdam, in press, 1998a.

Paulin, M.J., Phillips, R., Clark, J.I., Trigg, A., and Konuk, I., A full-scale investigation into pipeline/soil interaction. Proceedings of 1998 International Pipeline Conference, ASME, Calgary, AB, Canada, June 1998, vol. 2 pp. 779-787, 1998b.

Piplin-III, Computer program for stress and deformation analysis of pipelines. Structural Software Development Inc., California, 94704, 1991.

Pipeline Research Council International, Inc. (PRCI), "Enhancement of integrity assessment models/software for exposed and unburied pipelines in river channels", Danish Hydraulic Institute Report to PRCI, February 2000.

Pipeline Research Council International, Inc. (PRCI), "2000, Extended model for soil pipe interaction" Report prepared by C-CORE and D.G. Honegger Consulting, August 2003.

Pipeline Research Council International Project PR-268-9823. Honegger, D.G., and Nyman, J., Manual for the Seismic Design and Assessment of Natural Gas Transmission, 2001.

Pipeline Research Council International Project PR-268-9823.Honegger, D.G., and Nyman, J., Guidelines for the Seismic Design and Assessment of Natural Gas and Liquid Hydrocarbon, 2004.

Pipeline Research and United States Department of Transportation, Honegger D.G. et. al. (Draft proposed to be published in 2009), Guidelines for Constructing Natural Gas and Liquid Hydrocarbon Pipelines Through Areas Prone to Landslides and Subsidence.

Porter, M. et al.. Estimating the influence of natural hazards on pipeline risk and system reliability. In: 5th INTERNATIONAL PIPELINE CONFERENCE, 2004, Calgary, Alberta Canada,. Proceedings CD, 2004.

Porter, M., R. Reale, G. Marcuz, K.W. Savigny., "Geohazard risk management for the Nor Andino Gas Pipeline", Proceedings of IPC 2006, 6th International Pipeline Conference, September 25-29, 2006, Calgary, Alberta, Canada.

Porter, M., Savigny, K.W., Natural hazards and risk management for South American pipelines. In: 4th INTERNATIONAL PIPELINE CONFERENCE, 2002, Calgary, Alberta, Canada,. Proceedings: CD, 2002. 141.

Rajani, B.B.; Robertson, P.K.; Morgenstern, N.R., Simplified Design Methods for Pipelines Subject to Transverse and Longitudinal Soil Movements. Pipeline Technology – OMAE, Canada, Volume V, pp. 157-165, 1993.

Rizkalla, M. and McIntyre M.B., A Special Pipeline Design for Unstable Slopes, Energy Technology Conference and Exhibition (ETC&E), Houston, 1991.

Rizkalla, M., Poorooshasb, F., and Clark, J.I., Centrifuge Modelling of Lateral Pipeline/Soil Interaction. Proceedings of 11th Offshore Mechanics and Arctic Engineering Symposium, 13p., 1992.

Rizkalla M., Turner R.D. and Colquhoun I., State of Development of A Fiber Optic Technology-based Pipeline Structural Integrity Monitoring System , Proceedings of the 12th International Conferences on Offshore Mechanics and Arctic Engineering, 1993.

Rizkalla, M., A. Trigg and G. Simmonds, "Recent Advances in the Modeling of Longitudinal Pipeline/Soil Interaction for Cohesive Soils," Proceedings of the 15th International Conference on Offshore Mechanics and Arctic Engineering, American Society of Mechanical Engineers, Vol. V, pp. 325-332, 1996.

Rizkalla M. and Randell C., A Demonstration of Satellite-based Remote Sensing Methods for Ground Movement Monitoring in pipeline Integrity Management, Proceedings of the 18th International Conferences on Offshore Mechanics and Arctic Engineering, July 11-16 1999, St. John's Newfoundland, Canada, 1999.

Rizkalla M. and Read R., 2007, The Assessment and Management of Pipeline Geohazards, Rio Pipeline Conference & Exposition, 2007, Rio de Janeiro. Proceedings. Brazil.

Reed M., Personal communication on the application of geohazard data management systems in Alliance Pipeline, 2008.

Slope Indicator, Application Stories. http://www.slopeindicator.com/stories/pipeline.html.

Song, B., J.J.R. Cheng, Chan, D.H. and Zhou, J., "Numerical Simulation of Stress Relief Of Buried Pipeline at Pembina River Crossing", Proceedings of the International Pipeline Conference, Calgary, Alberta, Canada, 2006.

Sorensen, M.L., "Pipeline Maintenance in Geotechnically Unstable areas: A Case Study", Proceedings of the International Pipeline Conference, Calgary, Alberta, Canada, 2000.

Sutherby R., Fenyvesi L., Colquhoun I. and Rizkalla M., System wide Risk-based Pipeline Integrity Program at TransCanada, Proceedings of the International Pipeline Conference, Calgary, Alberta, 2000.

Sweeney, M., Gasca, A.H., Garcia Lopez, M., and Palmer, A.C. "Pipelines and landslides in rugged terrain: A database, historic risks and pipeline vulnerability. Proc. Terrain and Geohazard Challenges Facing Onshore Pipelines", Thomas Telford, London, 2004.

Sweeney, M., "Terrain and geohazard challenges facing onshore oil and gas pipelines: historic risks and modern responses," International Conference on Terrain and Geohazard Challenges Facing Onshore Oil and Gas Pipelines: Institution of Civil Engineers, London, UK, 2005.

Trigg, A.; Rizkalla, M., Development and Application of a Closed Form Technique for the Preliminary Assessment of Pipeline Integrity in Unstable Slopes. Pipeline Technology – OMAE, Canada, Volume V, pp. 127-139, 1994.

Tennyson, R.C., Morison, W.D., and Colpitts, B., Brown, A., "Application Of Brillouin Fiber Optic Sensors To Monitor Pipeline Integrity", Proceedings of the International Pipeline Conference, Calgary, Alberta, Canada, 2004.

Younden, J. et al.. Satellite-based monitoring of subsidence ground movement impacting pipeline integrity. In: 4th International Pipeline Conference, 2002, Calgary, Alberta, Canada, Proceedings: ASME, 2002, CD.

Zhou, J. et. al., "An emerging methodology of slope hazard assessment for natural gas pipelines", Proceedings of the ASME 2000 International Pipeline Conference, Calgary, Alberta, Canada. 2000a.

Zhou, J. et. al., "A methodology to maintain pipeline integrity at water crossings", Proceedings of the ASME 2000 International Pipeline Conference, Calgary, Alberta, Canada. 2000b.

Zimmerman, T., Nessim, M., McLamb, M., Rothwell, B., Zhou, J., and Glover, A. 2002, "Target reliability levels for onshore pipelines", Proc. 4th International Pipeline Conference, Calgary, AB, Canada, 2002.

Index

A

ABAQUS, 258
Adhesion factor, 261, 262
Aerial patrol surveillance (APS), 286-287, 288
Air photography interpretation, 288, 292
Airphoto imagery, acquisition of, 10-11
Airphoto remote sensing, 15-19
ALA (American Lifelines Alliance), 262
Alluvial fans, 103
American Lifelines Alliance (ALA), 262
American Society of Mechanical Engineers (ASME), 256
Anchor installation drill, 196
Anchor pull test, 194
Anchor weighting spacing, 217-221
Anchors, 188-199
corrosion of, 199
freeze-back, 197-199, 206
grouted, 195-197, 205-206
screw, 190-194, 204-205
APS (aerial patrol surveillance), 286-287, 288
ASME (American Society of Mechanical Engineers), 256
Attapulgite, 147
Axial soil springs, 260-261

B

Backfill, straw, 324
Backfill materials, manufactured, 324-325
Backfill shear resistance, 210-211
Bank erosion, 94-97
Bank restoration and armoring techniques, 124
Barrel reamer, 136
Bed restoration techniques, 125
Bending stress, 154-155

Bentonite, 147
Bolt-on weights, 178-180, 202-203
Buoyancy, 169-226, ii-iii
as-built drawings, 226
calculation input values, 209-212
control item locations work process, 221-224
control options, 174-206
control spacing and locations, 214-226
definitions and abbreviations, 170-172
design forces, 207-214
design philosophy, 173
installation, 225-226
operational temperatures and, 213-214
pipe stress and, 213
pipeline codes, 172
procurement process for, 224-225
safety factor, 212-213

C

Canadian Standards Association (CSA), 253
Catastrophic failure, 269
CCC (continuous concrete coating), 176-178, 201-202
Cellulose, polyanionic, 147
Channel changes, 97
Channel reinforcement, 316
CMC (sodium carboxymethyl cellulose), 147
Coastal terrains, 46-49
Columnar basalt, 46
Competent velocity method scour calculations, 92, 93
Concrete weighting spacing, 214-216
Concrete weights, 174-182
continuous concrete coating, 176-178, 201-202

dimensions, 181
reinforcement, 182
river weights, 178-180, 202-203
saddle weights, 174-176, 201
soil weights, see Soil weights
Consequence analysis, 254, 270
Continuous concrete coating (CCC),
176-178, 201-202
Conventional survey techniques, 290,
298-299
Creep deformation, 269
CSA (Canadian Standards Association),
253

D
Daywork contracts, 164
Debris flows, 103, 105, 106
Deep burial, 317-318
Deeper trench soil weights, 182-183, 203
DEM (Digital Elevation Model), 244, 245
Depth of cover, re-establishing, 316-317
Desert terrains, 53-57
Design, 76n
Design Flood Criteria, 73-76
Design steps and considerations, 76-80
Dewatering of excess drilling fluid,
150-151
Digital Elevation Model (DEM), 244, 245
Ditch modifications, pipeline, 326-327
Drag
fluidic, 152
frictional, 152
Drainage, 313-314
Drainage diversion berms, surface, 315
Drainage patterns, 30
Drilled path design, 140-142
Drilled path method, 161
Drilling fluids, 145-151
composition of, 146

description of components of, 147
excess, 149-151
flow schematic, 145
functions of, 145-146
inadvertent returns, 148-149
Dunes, 54-55

E
Earthquakes, 7
EIA (environmental impact assessment), 8
Elevated pipeline river crossings, open
cut and, see Open cut and elevated
pipeline river crossings
Environmental impact assessment (EIA), 8
Expert judgment, 268-269
Extensometers, 289, 298
External hoop stress, 155-156

F
Fault rupture, 7
Fiber optic sensors, 301, 307
Field data requirements for river crossings, 80-81, 82-86
Field inspection form, 251
FLAC, 258
Flood Criteria, Design, 73-76
Floodplain characteristics, 86
Flow isolation techniques
schematics of, 123
summer, 121
winter, 122
Fluidic drag, 152
Fluvial terrain, 28-33
Flycutter, 136
Freeze-back anchors, 197-199, 206
Frequency analysis, 254
Frequency Index, 278
Frictional drag, 152

G
Geographic Information Systems (GIS), 14
presenting pipeline terrain data in,

345

60-62
synthesizing terrain datasets within,
58-60
Geohazards, 6, 7
associated, global terrains and, see
Global terrains and associated geohazards
categories of, 235
consequence assessment, 270
considerations in pipeline risk management, 256-260
credible probable, 234
data management for, 244-253
describing, 243-244
design and operational mitigation,
308-328
importance of management of, 229-233
inputs to management, 328-330
inventory of, 235-243
management of, 229-334, iii
management plan, 330-334
management planning process, 328-334
monitoring, 285-308
monitoring ROWs, 286-299
aerial methods, 286-293
ground instrumentation, 289, 294-298
satellite-based methods, 288, 293-294, 295
support instrumentation, 289, 298-307
options for mitigating, 236
pipeline failure frequencies associated with, 230
pipeline integrity relating to, 232
pipeline monitoring summary, 302-303
pipeline protection, 319-323
planning of monitoring programs, 308
preliminary assessment of, 277
proven mitigation technologies, 310-313
ranking methodology for, 277-280

regional, 233-244
relevance of, 234
risk assessment, 267
risk assessment methodologies, 253-284
risk management in pipeline context,
253-256
semi-quantitative and qualitative assessment, 275-277
susceptibility assessment, 267-270
susceptibility mapping, 281-284
Geophysical methods, 299
Geotextile swamp weights (GSWs),
186-187, 203-204
GIS, see Geographic Information Systems
Glaciated terrains, 23-27
Glacier dammed lake releases, 103, 104
Global terrains and associated geohazards, 22-57
coastal terrains, 46-49
desert terrains, 53-57
fluvial terrain, 28-33
glaciated terrains, 23-27
karst terrains, 49-53
monitoring the pipe, 299, 300, 301
mountain terrains, 40-43
peatland/wetland terrains, 38-40
permafrost-affected terrain, 34-38
volcanic terrains, 43-46
GMSI (ground movement susceptibility index), 270
Granular bag weights, 183-186
Ground, avoiding unstable, 325
Ground improvement, 313-314
Ground movement susceptibility index
(GMSI), 270
Ground movements, 285-286
Ground survey, precise, 290, 298
Ground truthing, 20-21
Grouted anchors, 195-197, 205-

206
GSWs (geotextile swamp weights),
186-187, 203-204
Gullying, 7
Gypsum, 7

H
H/D ratio, 264, 265
HDD, see Horizontal directional drilling
HEC-6, 89
HEC-RAS, 77
Hole opener, 137
Horizontal directional drilling (HDD),
69, 109, 133-165, ii
construction monitoring, 164-165
contractual considerations, 163-164
drilled path design, 140-142
drilling fluids, 145-151
external coating, 163
installation loads, 151-153
installation stresses, 153-156
operating loads, 156-158
operating stresses, 158-160
pipe specification, 151-163
process of, 133-139
pulling load calculation by the PRCI
Method, 160-162
site investigation, 139-140
temporary workspace requirements,
142-145
Horizontal drilling rig, 142-143
Horizontal soil springs, 262-264

I
IL (level index), 276
ILI (inertial in-line inspection), 301,
304-306
Imported fill, 183
Inertial in-line inspection (ILI), 301,
304-306
Initiation Index, 278
InSAR (interferometry using radar
imagery), 288, 293
Installation stresses, 153-156
Instream and parallel pipeline alignment, 107, 108
Interferometry using radar imagery
(InSAR), 288, 293

K
Karst, 7
Karst terrains, 49-53

L
Land farming, 150
Landscape changes, 17
Landslides, 7
Lava flows, 44
Level index (IL), 276
LiDAR (Light Detection and Ranging)
imagery, 11, 245
LiDAR survey, 288, 292-293
Light Detection and Ranging, see LiDAR
entries
Limestone, 7
Liquefaction, 7
Lump sum contracts, 163

M
Manufactured backfill materials, 324-325
Map remote sensing, 12-13
Maps, acquisition of, 10-11
Mineral soil density, 209-210
Monitoring requirements for river crossings, 126, 128, 129-130
Mountain terrains, 40-43
MSS (multispectral scanners), 14
Multispectral scanners (MSS), 14

O
Open cut and elevated pipeline river
crossings, 69-130, ii
approach to, 69-70
construction, 111-124

design, 70-111
operations, 126-130
Operational temperatures,
buoyancy and,
213-214
Optical imagery, 288, 294, 295
Organic soil density, 210
Overfill and post construction
settlement, 127

P
Peatland/wetland terrains, 38-40
Permafrost, 171
Permafrost-affected terrain, 34-
38
P.I. elevation, 141
Pilot hole, 133-137
Pipe excavation for strain relief,
327
Pipe isolation, 317
Pipe monitoring techniques,
other, 301,
307
Pipe stress, buoyancy and, 213
Pipe wrap, synthetic, 324
Pipeline ditch modifications, 326-
327
Pipeline failure frequencies
associated
with geohazards, 230
Pipeline integrity relating to
geohazards, 232
Pipeline mitigation, 317
Pipeline Research Council
International
(PRCI) Method, 160-162
Pipeline river crossings,
elevated, open
cut and, see Open cut and
elevated
pipeline river crossings
Pipeline route selection and
characterization, 5-62, ii
airphoto remote sensing, 15-19
integrating multidiscipline study
data, 8, 9
map remote sensing, 12-13
process of, 5-7
remote sensing from maps,
satellite

imagery and airphotos, 9-22
satellite imagery remote sensing,
14
Pipeline terrain data, presenting,
in
Geographic Information
Systems,
60-62
PIPLIN, 258-260
Plate weights (PWs), 187-188,
203-204
Polyanionic cellulose, 147
Polymers, 147
PRCI (Pipeline Research
Council
International) Method, 160-162
Precise ground survey, 290, 298
Prereaming pilot holes, 135-137
Pull section breakover, 138
Pull section fabrication, 143-144
Pullback, 137
Pulling load calculation by the
PRCI
Method, 160-162
PWs (plate weights), 187-188,
203-204

Q
Quantitative risk assessment
(QRA), 267,
270
Quantitative versus qualitative
analysis, 71-72

R
Radius of curvature, 141-142
Rainfall-slope movement model,
272
Rate Index, 278
Regional geohazards, 233-244
Remote sensing from maps,
satellite
imagery and airphotos, 9-22
Retaining walls, 315
Right-of-ways (ROWs), 17
monitoring geohazards for, see
Geohazards, monitoring ROWs
"Ring of Fire," 43
Risk analysis, 253-254
Risk estimation, 254

Risk evaluation, 253
Risk matrix methods, 254-255
River channel migration, 7
River classification system, 73
River crossings
arctic, 128
buried, 78
elevated, 79, 109, 111, 112-115
open cut and, see Open cut and
elevated pipeline river
crossings
field data requirements for, 80-
81,
82-86
girder bridge, 113
intermediate, 99, 117
major, 101, 102, 116
major buried, 98
monitoring requirements for, 126,
128,
129-130
open cut
with flow diversion, 118
in winter, 120
short span elevated, 112
suspension bridge, 114
typical, 100
River flow, 80, 87
River training structures, 107,
109, 110
River water level, 87-89
River weights (RW), 178-180,
202-203
Riverbank, see Bank entries
Riverbed scour, 33, 89-94
Route data, availability of, 253
ROWs, see Right-of-ways
RW (river weights), 178-180,
202-203

S
S (susceptibility rank), 278
Saddle weights (SW), 174-176,
201
Sand seas, 53
Satellite imagery, acquisition of,
10-11
Satellite imagery remote
sensing, 14
Satellite sensor types, 11

Scour, riverbed, 33, 89-94
Scour Hazard Database Model
(SHDM), 272,
273
Scour multiplication factors, 96
Screw anchors, 190-194, 204-
205
Seismograph, 289
Semi-quantitative risk
assessment
(SQRA), 275-276
SHDM (Scour Hazard Database
Model), 272,
273
Sinkholes, 49-53
Slope grading, 315
Slope inclinometers, 289, 294-
296
Slope movement hazard
analysis, 271
Sodium carboxymethyl cellulose
(CMC),
147
Soil densification or
improvement, 314
Soil erosion, 7
Soil liquefaction, 213
Soil springs
axial, 260-261
horizontal, 262-264
vertical bearing, 266
vertical uplift, 264-266
Soil weighting, 216-217
Soil weights, 182-288
deeper trench, 182-183, 203
SONDEX tubes, 289, 296, 297
SQRA (semi-quantitative risk
assessment), 275-276
Stability analysis, 269
Stereoscopic airphotos, 11
Strain gauge instrumentation,
299, 300,
301
Strain relief, pipe excavation for,
327
Straw backfill, 324
Stream channel types, 28-29
Strip mosaics, 17
Subsurface data, 245
Subsurface site investigation, 21-

22
Surface drainage diversion
berms, 315
Survey techniques, conventional,
290,
298-299
Susceptibility, 257-258
Susceptibility mapping, 281-284
Susceptibility rank (S), 278
SW (saddle weights), 174-176,
201
Swamp weights, 174-176, 201
geotextile, 186-187, 203-204
Synthetic pipe wrap, 324

T
Tensile stress, 154
Terrain datasets, synthesizing,
within
Geographic Information
Systems,
58-60
Thaw depressions, 36
Thermal expansion, 157
Thermal regime, maintenance of,
314
Thermokarst, 36
Till, 23
Toe berm, stabilizing, 315
Topographic survey, 139-140
Triggering events, 268

U
Underwater diver survey, 290,
298

V
Vegetation, 314
Vertical bearing soil springs, 266
Vertical uplift soil springs, 264-
266
Volcanic terrains, 43-46
Volcanoes, 7
Vulnerability, 257
Vulnerability criteria, 258-259
Vulnerability Index, 278

W
Wetland terrains, 38-40
Wood lagging, 179, 180

Workspace requirements,
temporary,
142-145

X
Xanthan gum, 147

Z
Z-factors, 96

About the Editor

Moness Rizkalla, M. Eng. P. Eng.

Moness Rizkalla is a recognized specialist in pipeline design and integrity management. His experience includes several management and senior technical roles in the areas of pipeline design, project management, pipeline risk assessment and integrity management planning, operations support and associated technology development and application. Moness' professional experience has been gained in both Canadian and international settings for both pipeline operators and consulting firms.

Within the pipeline integrity management arena, he has specialized in the management of external load hazards – geotechnical and mechanical damage. He has been involved in a wide range of pipeline geotechnics with an extensive list of associated publications. In 2000, Moness founded Visitless Integrity Assessment Ltd., a company that delivers commercial proactive prevention solutions to the pipeline industry's mechanical damage challenges.

About the Authors

Chapter 1: Introduction
Moness Rizkalla M. Eng. P. Eng.
Please see About the Editor.

Chapter 2: Pipeline Route Selection and Characterization
J. D. (Jack) Mollard, OC, PhD, LLD, P. Eng., P. Geo.
Jack Mollard is founder and president of a firm of consulting engineers and geoscientists specializing in satellite imagery interpretation and application. He is responsible for locating tens of thousands of kilometres of proposed and built pipeline routes while mapping and avoiding geohazards during right-of-way terrain mapping. He is also responsible for locating oil and gas pipelines in northern Canada's permafrost terrains including the proposed Mackenzie Valley and Polar Gas pipelines. He has also located and evaluated pipeline routes in the USA, Asia, Europe, South America and Australia.

Jack holds a MSSC from Purdue University in 1947 and a PhD from Cornell in 1952 and is the recipient of numerous national and international awards, honours, recognitions in Canada in the United States.

L. A. (Lynden) Penner, MSc, P. Eng., P. Geo.
Lynden Penner is a senior geotechnical engineer at J. D. Mollard and Associates with over 20 years consulting experience. He has extensive experience in pipeline route selection and terrain mapping across western and northern Canada. His experience also includes mapping pipeline

geohazards and research on detecting, predicting and parameters causing stress corrosion cracking on buried natural gas pipelines. He holds an MSC degree in geological engineering from the University of Saskatchewan.

T. A. (Troy) Zimmer, BSc, Hons Dpl, MCRSS

Troy Zimmer is a senior remote sensing and GIS specialist at J. D. Mollard and Associates with ten years of experience. He specializes in integrating digital remote sensing data and with GIS data in support of engineering and environmental applications, including computerized 3D virtual variety flyovers of proposed and built pipeline routes. Troy holds an Honours Diploma in Biological Sciences from the Northern Alberta Institute of Technology and a BSc in combined Biology / Geography from the University of Regina, Saskatchewan.

Chapter 3: Open Cut and Elevated Pipeline River Crossings
Wim M. Veldman, FEIC., P.Eng., B.Sc. (Civil Engineering), M.Sc. (Civil Engineering – Water Resources)

Wim has had a leading role in the design, construction and operational monitoring of river crossings for numerous major pipelines in North and South America. Long term consultancy to projects such as the Trans Alaska oil pipeline (35 years) and the Gas Atacama pipeline in Argentina (10 years) – technical, topographic and land use issues required numerous river crossings, floodplain segments and protective structures for both pipelines – provide the sound basis for lessons learned and the recognition of what is and what is not important in the design of river crossings. Wim has published papers on pipeline river crossings in ASCE and ASME proceedings and has testified on river crossings at regulatory hearings. His engineering firm has been awarded provincial and national awards for its pipeline work.

Chapter 4: Horizontal Directional Drilling
John D. Hair, P.E.

President, J. D. Hair & Associates, Inc.
John Hair is a recognized expert in the field of horizontal directional drilling (HDD) having been intimately involved with the development and application of HDD technology since 1980. He is the author of numerous technical papers and presentations on HDD and is regularly called upon to testify before regulatory bodies relative to the impact of HDD construction methods. Mr. Hair is a 1974 civil engineering graduate of Louisiana State University and is licensed to practice engineering in thirty three states.

Jeffrey S. Puckett, P.E.

Project Manager, J. D. Hair & Associates, Inc.
Jeff Puckett has been actively involved in the application of horizontal directional drilling (HDD) since 1994. As an engineer with J. D. Hair & Associates, he has provided design and construction management services on numerous projects involving the installation of pipelines by HDD.

Mr. Puckett graduated from the University of Oklahoma in 1992 with a Bachelor's Degree in Mechanical Engineering and is a licensed Professional Engineer in the State of Oklahoma.

Chapter 5: Overland Pipeline Buoyancy Control
Ray Boivin, M. Eng. P. Eng.

Ray Boivin is a geotechnical design engineer who started out in the consulting industry, then worked on the owner's side for 12 years, and now is back on the consulting side with his company Plateau Geotechnical Ltd. Ray has 26 years experience with 18 years in the natural gas and liquids pipeline sectors. Most of his work has been in western and northern Canada but work experience also includes Argentinean and Mexican transmission pipelines. Areas of experience and specialization include: pipeline route selection and optimization, watercourse crossing assessments and design (open cut and HDD), pipeline buoyancy assessment, geotechnical site investigations and regulatory applications and hearing support. Ray makes his home in Calgary, Canada with his wife and two daughters, the eldest of which is pursuing her engineering degree.

Chapter 6: Geohazard Management
Moness Rizkalla, M. Eng. P. Eng.

Please see About the Editor.

Rodney Stewart Read, Ph.D, P. Eng., P.Geol

Dr. Read is a practicing consulting engineer, specializing in applied rock mechanics and geotechnical engineering. He is President of RSRead Consulting Inc., a Canadian consulting firm based in Okotoks, Alberta, Canada. He has been involved in projects ranging from concepts for nuclear waste disposal to geohazard assessment for pipelines in challenging physiographic environments. Recently, he was geotechnical lead on the Turtle Mountain Monitoring Project at the historic Frank Slide in southern Alberta, Canada - the site of Canada's deadliest landslide. His current interests involve geotechnical risk assessment of linear systems in difficult terrain.

Gregg O'Neil, M.Eng. P.Eng.

Mr. O'Neil is currently a Principal with Klohn Crippen Berger Ltd. in Calgary, responsible for all oil and gas projects in the company. Mr. O'Neil's pipeline background includes design and managerial positions with NOVA Gas Transmission, TransCanada Pipelines and Foothills Pipelines. At the design level, Mr. O'Neil has been involved primarily in the geotechnical components of pipeline design and pipeline integrity projects.